TE DIE

BOSTONIANS AND BULLION

BOSTONIANS
AND BULLION

The Journal of Robert Livermore
1892 – 1915

Livermore, Robert

Edited by
GENE M. GRESSLEY

UNIVERSITY OF NEBRASKA PRESS · LINCOLN

MANUFACTURED IN THE UNITED STATES OF AMERICA

Contents

A section of illustrations follows page 92.

Maps follow pages 40, 82, 98, 120, and 166.

Introduction

One of the most toasted of all American mining engineers, John Hays Hammond, in his none too modest autobiography, recalled, "The mining investor is constantly obliged to take risks, even when prospects seem most favorable."[1] Another mining engineer turned author and editor, Thomas Rickard, in his memoirs insisted, "Mining is an adventure, a sane adventure, it is rarely an investment, and it need not be a gamble. It is a reasonable speculation."[2] Whatever sentiments seem real—"speculation" or "sane adventure"—the mining industry was a gamble, perhaps not as much of a gamble as playing faro at the National Club in Telluride, Colorado, but for some investors the gaming table would have been a more pleasant way to loose their gold pieces.

In this turn-of-the-century milieu of chance, one of the most intriguing and historically neglected figures was the mining engineer.[3] Sent by the corporation or lone investor to the remote regions of the world, he gave advice on the strength of which investments in the millions were made—and lost. His knowledge, experience, and—some mining-camp wags would say—intuition were all brought to bear to hedge the degree of speculation. In retrospect, what appears so remarkable is not the number of poor judgments, but the numerous correct ones.

[1] John Hays Hammond, *The Autobiography of John Hays Hammond*, 2 vols. (New York: Farrar & Rinehart, 1935), p. 513.

[2] Thomas A. Rickard, *Retrospect* (New York: McGraw-Hill, 1937), p. 52.

[3] Professor Clark Spence of the University of Illinois is undertaking a history of mining engineers. See his excellent initial treatment "The Mining Engineers in the West," in *The American West* (Santa Fe: Museum of New Mexico Press, 1963), pp. 100–110.

As a science, mining engineering and geology were in their rudimentary stage. One outstanding American mining engineer, William M. Fitzhugh, described the mining engineering curriculum as "seetee"—mechanical, civil, and general engineering.[4] His remark was a commentary on the state of education of engineers; not until after the beginning of the twentieth century did American universities achieve the distinction of their European counterparts, the Royal School of Mines in London or the Ecole des Mines in Paris, in the training of mining engineers.

After the classroom, a neophyte engineer with the proper connections might apprentice as an assistant engineer in the famous Rand district in South Africa or the Camp Bird mine in the San Juans of Colorado. More commonly, he associated with a state or national geological survey, as did young Richard Penrose, fresh from Harvard. If none of these fortunate alternatives were presented, he struck out on his own.

Many a young engineer had second thoughts about his chosen profession after experiencing a few months on his first job. Isolation in the bleak wastes of Alaska or having to endure the stultifying heat of South Africa or the scorpions in Mexico seemed a high price to pay for an occupation, especially when homesickness struck. Nor were the climatic conditions the only vexations; the brawling, raucous life in the mining towns kept life hazardous if interesting.

Eben Olcott, an engineer-manager residing in a small mining camp in Panama, in 1878 reminisced to his sister:

> I am getting to be looked on as a patriarch here. A man who has been here two years is considered an old inhabitant. I can certainly look back and note many improvements and changes. We seem to be getting more civilized. The shops look more citified, the houses are thicker, mule cars and wagons begin to take the place of pack animals, expensive labor saving machinery begins to take the place of anti-diluvium [sic] tools, assaying and chemical tests are dreamt of in lieu of pan washings, but unhappily the world don't get any better. As many men get drunk, as much dishonesty is practiced, immorality still abounds and each new mining company only seems to bring its increment and add to the mass.[5]

[4] Interview, William M. Fitzhugh, Jr., May 31, 1962.
[5] Eben Olcott to Pheme Olcott, June 19, 1878, Olcott Collection, Western History Research Center, University of Wyoming.

In one paragraph, Eben Olcott stopped time to describe the veneer of civilization which was slowly creeping over his region of Panama. Many another mining engineer in another place, at another time, made similar observations with much of the same sardonic attitude. Sanity required the mining engineer to be somewhat philosophical. Loneliness, disease, inclement weather—these were all ingredients of a mining engineer's life.

In addition to environmental conditions, there was always the plague of human relations in a strange culture. More than one engineer gave up in disgust, not because the mine played out, but because of his inability to manage the miners. Distance alone created problems, particularly with one's investor two thousand miles away, and several weeks by mail. Even when a mining engineer-manager was financially involved in a corporation, difficulties arose. J. R. Robinson, writing to his fellow Bostonian and investor, Charles Varney, concerning their mine in the Santa Eulalia complex in Chihuahua, Mexico, gnawed, "I do not feel that it is just or right or proper for me to consent to join the proceedings to foreclose the company. The plan you propose for the sale of the organization does not strike me as altogether just." After rambling on for a page, Robinson conceded, "It is possible that my head is somewhat mixed as to the plan." [6]

Nor did being relatives alleviate the tensions of distance and poor communication. Henry Seligman, of the brokerage firm of Seligman and Sons in New York, wrote his nephew, Albert Seligman, in Montana, "I have no doubt but the proposition you made is a very fair one and will turn out a profitable investment. But we cannot afford to block up any more cash as we require all of it to take care of our railroad properties, which as you are well aware are not in a very flourishing condition." [7] In spite of angry words and pleas from his nephew, Henry Seligman refused to budge. What annoyed Albert most was the thought that Henry seemed to imply that Albert had poor business sense. "All those questions you ask, have common sense answers and you know it," retorted Albert. [8]

[6] J. R. Robinson to Charles T. Varney, February 24, 1887, Robinson Collection, Baker Library, Harvard School of Business Administration.

[7] Henry Seligman to Albert Seligman, January 17, 1885, Seligman Collection, *ibid.*

[8] Albert Seligman to Henry Seligman, April 22, 1887, *ibid.*

Some engineers would have willingly settled for altercations with their investors as the easiest path to follow in a perilous occupation. For those engineers who misjudged the potential of a prospective property, advising their investors to gamble and then discovering that their examination erred could make life a veritable hell. Solid reputations built up over the years could vanish overnight. The mining engineering fraternity was a close one; once a *faux pas* had been made, it was soon common knowledge throughout the mining world. In fact, a famous miscalculation could be debated in mining conventions twenty years after it occurred.[9]

Thomas Rickard, whose British genealogy was studded with mining engineers, retreated to the security of journalism after two unfortunate examinations in as many years.[10] The first embarrassment was with the Independence mine in Cripple Creek, and then followed one with the Camp Bird near Ouray, Colorado. In the latter assessment he valued the property at $6,000,000. His British clients, the Venture Corporation of London, asked John Hays Hammond to make a corollary estimate. Hammond, accompanied by Hennen Jennings, whose fame in the Rand in South Africa was widespread, made the examination, coming up with the conclusion that Rickard's appraisal was just double their findings. This led Hammond to make the stinging comment, "In spite of their reputation [they] seem to lack the proper qualifications for evaluating mines."[11]

The incident had a bitter aftertaste for Rickard when the next year his brother-in-law, A. Chester Beatty, was nominated to confirm Hammond's judgment. As for Rickard, he was only too happy to escape the censure of his peers by assuming the editorship of the *Engineering and Mining Journal* in the summer of 1902. In his autobiography, three decades later, Rickard was still full of bitterness and embarrassment when he recalled the Camp Bird incident:

> So long as I examined relatively small mines for individuals, my work as a consulting engineer was delightful. . . . But when I was

9 Thomas A. Rickard, *Retrospect* (McGraw-Hill: New York, 1937), p. 76.
10 *Ibid.*, pp. 75–82.
11 While Hammond's caustic observation on Rickard had some justice, it was more than a bit arrogant. Hammond later confessed that his report on the Nipissing silver mine in Canada proved inaccurate, resulting in substantial losses to the Guggenheims. He was, of course, misled by the miscalculations of others, but he nobly assumed sole responsibility (*ibid.*, pp. 482, 515).

called to report upon mines the purchase of which involved millions of dollars, I found that I had to be mixed up in a kind of business for which I had a decided distaste. One can not pick one's clients and one is compelled to face realities.[12]

A mining engineer's life was not all physical discomfort, social privation, rotting disease, and inevitable risk; weighted against these negative aspects there was adventure, excitement, and opportunity associated with international society, the challenge to match one's wits against both man and nature, and, above all, the satisfaction of professional accomplishment. Eben Olcott believed that the field of mining engineering did offer a young man an open-ended future; he wrote to his sister, "Sometimes I have wished that I had selected a different occupation, but then look what I would have missed. There are more wonderful satisfactions from being right than there ever are disappointments in making the wrong decision."[13] Eben was obviously feeling buoyant on that beautiful winter day in the Andes, for a few weeks before, his optimism was decidedly muted. Of course, for many engineers it was much easier to recall the pleasant aspects of one's life when cushioned in a chair in one's Cape Cod home, than it was thirty years before, trudging along the spine of a mountain.

Most mining engineers agreed with Olcott's romantic sentiments. One of the striking, though perhaps expected, undertones in Rickard's rather staid and unimaginative *Interviews with Mining Engineers* was the recurring observation that "if I had to do life over again, I would trod the same path." After all, though, who at retirement age was about to admit life had been a failure?

Certainly, Robert Livermore considered his life one of full adventure. "I have had a lot of fun out of life," was the first thought that occurred to him when he sat down to begin his autobiography. Although readers are prone to discount such exuberant reflections as clichés, Robert Livermore meant every word. He did enjoy life—and to the fullest!

The world of Boston that greeted Robert Livermore in 1876 was a stable one, particularly if one's name ended in Cabot or Lowell. Although the social extremes of a Frank Skeffington and a George

[12] Rickard, *Retrospect*, p. 82.
[13] Eben Olcott to Pheme Olcott, November 20, 1878, Olcott Collection, Western History Research Center, University of Wyoming.

Apley were real enough, they did not match the caricatures described by writers of best-selling fiction. "Proper Bostonians," for all their snobbishness and glacial attitudes, were neither so frigid nor so hopelessly lost to reality as they have often been portrayed.

The upper-class Bostonian was indeed proper in many ways, but he was not the Neanderthal specimen, who should be placed on display in a museum of natural history. He was very human, much like any of his fellow man, whether he was in the Somerset Club or in the counting house. This is not to say that Boston society was the same as that of New York, Philadelphia, or Chicago; it emphatically was not! The Bostonian scorned much of the life that a New Yorker found appealing. First and foremost, he was upper-middle-class in outlook and seldom extremely wealthy or even interested in the pursuit of riches for their own sake. More often than not, as Samuel Eliot Morison has written, great-grandfather had established the family fortune. Subsequent generations were assigned one economic task—to preserve it. To this family obligation, generation after generation held true—as is testified to by the innumerable family trusts in Boston today. This did not mean that they scorned gambling with the family resources, for Bostonians invested in as many risk-taking ventures as any other American,[14] but more often than not they speculated with the income, and not the principal of their estates.

With a relatively stable income, and freed from the family dictum to equal grandfather's acquisitive feats, the Bostonian of the Victorian age turned to scholarship, education, literature, and the pursuit of a cultural life. It was a world of leisure, of four horses displayed in front of the carriages, rather than three hundred concealed under the hood. Every family of means had horses, and a respectable lady did not hire transportation. It was an era of gaslights, well-stocked libraries, and even better stocked wine cellars. Social life revolved around dining out, not at the Ritz, but at each other's brownstone homes. These informal occasions were interspersed with formal teas and receptions.

Undoubtedly, many a visiting New York financier found the social customs of Boston quaint and even staid, when contrasted with the

[14] Many examples of the Bostonian's speculative spirit are found in Arthur Johnson's monograph, *Boston Capitalists and Western Railroad* (Cambridge: Harvard University Press, 1967), and Gene M. Gressley, *Bankers and Cattlemen* (New York: Alfred A. Knopf, 1966).

chaos of Delmonicos at high noon. True, it may not have been exciting and stimulating, still it was cultivated, serene, and, above all, comfortable. Other generations, in another time, could find much to envy without losing their perspective in nostalgia.

There was a strict tradition that dominated, a well-defined way of life, which historian Morison insists was indigenous to New England, and not an imported English facade. This may have been so, but it is hardly demeaning to the first families of Boston to suggest that they had much in common with the Tory aristocracy of England. Nor does it stretch credibility to suggest that perhaps a substantial English heritage was not shed the minute the first settlers stepped on the Massachusetts shore.

Definitely Boston could be intolerant, but its widely reputed provincialism was much more tempered by an educational base than has commonly been admitted. Some of the most liberal theological and philosophical thinking of the nineteenth century was "born" in Boston environs.

This then, was Robert Livermore's Boston. As he said, the Livermores were of good stock, but they were not "old family." His father, Colonel Thomas L. Livermore, was a vice president of the Calumet and Hecla Copper Company in Michigan and an investor in a host of mining companies.[15] Owned primarily by Bostonians—the Shaw, Agassiz, and Livermore families—the Calumet and Hecla, one of the richest bonanzas in all American mining, poured $155,000,000 in dividends into its stockholders' bank accounts between 1871 and 1923.[16]

One cliché avers that all babies born in Boston soon face the sea. Whether or not young Bob Livermore had an inborn inclination toward the sea is for speculation, but any young boy growing up in Boston in the nineteenth century found it difficult to escape the influence of generations of sea-faring New Englanders. A nautical tradition was as encrusted on Boston culture as barnacles on the bottom of ships' keels.

Coupled with Bob Livermore's love of the sea was a wanderlust,

[15] An adequate history of the Calumet and Hecla is William B. Gates, Jr., *Michigan Copper and Boston Dollars* (Cambridge: Harvard University Press, 1951).

[16] Charles W. Henderson, *Mining in Colorado* (Washington: Government Printing Office, 1926), p. 73.

and he left home at the age of seventeen to sail on the "tidy" training ship *Enterprise*. For the next eight years young Livermore blended adventure as a sailor, cowboy, and miner with acquiring an education at Harvard and Massachusetts Institute of Technology. Between his sophomore and junior years at Harvard, his father sent him west, in part to forestall his romantic notions of joining the army in the Spanish-American War. His father succeeded; after one trip, Robert Livermore was enamored of the West: "My taste of the rougher fringe of the West had fired my imagination to the point of trying more unsupervised contact with it." This desire introduced his cowboy interlude, which, from the standpoint of physical and economic comforts, was one of the leaner periods in his life. Although he made no open confession, one suspects that he returned to college with considerable relief.

In June, 1903, Robert Livermore finished his course in mining engineering at Massachusetts Institute of Technology. Very soon afterward he began his mining career as assistant engineer for the Camp Bird mine near Ouray, Colorado;[17] and intermittently for the next three decades he was associated with the mining industry in San Miguel and San Juan counties in central Colorado.

A year before the turn of the century, Thomas Livermore had purchased stock in the Smuggler-Union mine in Telluride. After the manager, Arthur Collins, was assassinated in a labor conflict in 1902, Colonel Livermore sent his dashing son-in-law, Bulkeley Wells, to assume the management. This made the decision all the easier for young Livermore when John Hays Hammond offered him a position at Camp Bird in 1903,[18] for Telluride and his family were just over the San Juan range. After a short sojourn in the Camp Bird, Livermore, with his lifelong friend, Halstead Lindsley, obtained a lease on the Smuggler-Union in 1904. There he worked in close association with Bulkeley Wells.

Wells was one of the most glamorous and controversial mining promoters and managers in the entire West. Born in Chicago on March 10, 1872, he graduated from Harvard in 1894.[19] A year later he married Grace Daniels Livermore. First employed in a manufac-

---

[17] For a family account of the history of Camp Bird, Evalyn Walsh McLean, *Father Struck It Rich* (Boston: Little Brown, 1936) is fascinating.

[18] Hammond's association with the Camp Bird is discussed in Hammond's *Autobiography*.

[19] Denver *Post*, May 26, 1931.

turing company in New Hampshire, he later secured a position on
the Boston and Maine Railroad. By 1896, he was in the West looking
after the widely scattered Livermore interests. Extremely handsome
and full of personal charm and courage, Bulkeley Wells was soon
dinner conversation in Telluride as well as elsewhere in Colorado.

To this day if one lunches in the gracious dining room atop the
Denver's Club skyscraper, and mentions the name of Bulkeley
Wells, the tinkle of ice will stop, and one of the party will set down
his glass and tell his favorite Wells story. For the tales about him are
myriad, and the variations legion.[20] Many of them are apocryphal,
for separating fact from fiction about Wells is as difficult as finding a
mother lode in Death Valley.

Many of his closest associates found him inscrutable, an enigma-
tic man with a multitude of facets to his personality. Some knew him
as a dashing promoter who secured the confidence of Harry Payne
Whitney and an entree to the Whitney millions. There is no question
that Wells invested some of the Whitney fortune; Harry Payne
Whitney's name appears on several of Wells' favorite mining stocks.
That he "guided" seventy-five million of Whitney's dollars, as is
frequently repeated, is a wild exaggeration.

To some he was an affable host and a charming guest who was
equally at home in Denver's finest society or on a picnic in the San
Juan Mountains, although the Wells-sponsored outings in the
mountains could hardly be classified as roughing it. Grace Livermore Wells recalled the mountain picnics as extremely elaborate
excursions, complete with fine china, vintage wines and linen
tablecloths.[21] If called upon, Wells could produce a jellied consommé
chilled in a mountain stream or deliver a succulent roast of wild
game basted with a carefully selected wine. One mining engineer
acquaintance of Wells remembered an incident illustrative of Wells'
epicurean nature that took place over fifty years ago. Meeting him on
a platform of a small Colorado railroad station, Wells invited the
friend to accompany him in his private car on the tiny narrow-gauge
railroad. As they were slowly wending their way along the pinched-in

[20] Biographical data on Wells is available in E. B. Adams, *My Association
with a Glamorous Man . . . Bulkeley Wells* (Grand Junction, Colorado: privately
printed, 1961); Denver *Post*, May 27, 1931; W. F. Stone, ed., *Supplement to the
History of Colorado* (Chicago: S. J. Clarke Publishing Co., 1918), pp. 257–258;
Rocky Mountain News, May 27, 1931; and Wilson Rockwell, "The Telluride
Story," Telluride *Times*, March 5, 12, and 19, 1965.

[21] Interview, Grace Livermore Wells, November 6, 1964.

mountain valley, Wells suddenly pulled the bell cord to summon the conductor. When the conductor arrived, Wells ordered him to stop the train immediately. He then climbed down from his car and walked to the bank of a little mountain stream, where he gathered some wild mint. Returning to the car, he proceeded to mix a round of mint juleps for the entire party.

To some, Wells was a man of reckless courage. During the Telluride labor violence, he debonairly dismissed the possibility of physical harm, even after a bomb blew up part of his home. His contemporaries frequently labeled his regime as a mine operator at the Smuggler-Union as inefficient. Yet the few records that remain of the companies with which Wells was associated, including the Smuggler-Union, show that several of them paid dividends. We will never know whether in the final balance sheet the Smuggler-Union paid the investors, although we do know that in the 1920's there was an increasing struggle to keep the mine free of bankruptcy. A good conjecture would be that the Smuggler did not reward the investors, but this fact hardly distinguishes it from dozens of other mining corporations scattered the length of the Rockies. As one turn-of-the-century mining engineer told the writer, "If I were investing in mining companies fifty years ago, I would much rather have the capital that went into the ground, instead of the returns that came out!" [22]

Understandably, Wells' extravagant living gave credence to the rumors of a depleted Smuggler-Union treasury. Few would argue that more prudent management would have resulted in higher dividends. However, as Robert Livermore noted in the twenties, internal management was only one of the problems in the game of making a stockholder happy.

Wells was divorced from Grace Livermore in 1918. He stayed on in Telluride and Denver another five years, dabbling in several other ventures—an electric power company, an irrigation scheme, and a radium mine. The last conceivably could have been a stroke of genius had not richer mines been discovered in the Congo. A drop in the world market of radium forced the Paradox mines into an uncompetitive position.

When Wells left Colorado in 1923, he bought, with Whitney's backing, control of the famous United Comstock mining complex near Virginia City, Nevada. He reopened the mine and constructed

22 Henry C. Morris to Gene M. Gressley, August 18, 1965.

a cyanide mill, but the mine never came close to making a profit. At Gold Cañon, not far from Silver City, Nevada, Wells started a gold-dredging operation, another unrewarding investment. An early-day mine, the Idaho-Maryland at Grass Valley, failed to yield to the Wells touch. He had better success with a few properties in Trinity County, California. When the price of silver plummeted in the early 1920's, the United Comstock collapsed. With its failure, the Whitney pocket book snapped shut, and the Wells fortune continued to dissipate. On May 24, 1931, he ended his life in his office in San Francisco.

With a few deft strokes by a first-rate scenario composer combination, it would be relatively easy to turn the life of Bulkeley Wells into an operatic melodrama, in the same superficial genre of the filmy, light summery take-offs, as the "Unsinkable Molly Brown," or the "Ballad of Baby Doe."

The social attributes of Bulkeley Wells aside, Livermore looked forward to transferring from Ouray to Telluride. Not only would he be among relatives and friends, but his economic opportunity in owning a paying lease was much better than that of the job at Camp Bird. When Bob Livermore first saw Telluride in 1904, it had 2,446 inhabitants.[23] Like that of all mining communities, Telluride's population fluctuated with the mercurial temperaments of the miners, the rumors of rich strikes elsewhere (the next mountain range always promised more), and the investment tendencies of eastern financiers. When Livermore arrived in Telluride, labor unrest brought on by the agitation of the Western Federation of Miners formed a large question mark on the town's future.[24]

[23] Twelfth Census of the United States, *Population* (Washington: Government Printing Office, 1901), p. 440.

[24] General background on the labor troubles in the Rockies can be found in: P. F. Brissenden, *The I.W.W., A Study of American Syndicalism* (New York: Macmillan, 1913); R. Chaplin, *Wobbly* (Chicago: University of Chicago Press, 1948); J. S. Gambs, *Decline of the I.W.W.* (New York: Columbia University Press, 1932); S. Perlman and P. Taft, *History of Labor in the United States, 1896–1932* (New York: Macmillan, 1935); Clarence Darrow, *The Story of My Life* (New York: Charles Scribner's Sons, 1934); Elmer Ellis, *Henry Moore Teller* (Caldwell, Idaho: Caxton Printers, 1941); J. K. Howard, *Montana, High, Wide and Handsome* (New Haven: Yale University Press, 1946); T. A. Rickard, *The History of American Mining* (New York: McGraw-Hill, 1932); and John A. Remington, "Violence in Labor Disputes" (Master's thesis, University of Wyoming, 1965).

The issues at stake centered on the eight-hour day and the termination of the fathom system. The fathom system, an innovation introduced into the San Juan region from the Cornish mines by Thomas A. Rickard,[25] was an arrangement whereby the miners were given a fixed price for each fathom that was completely removed. A fathom measured six feet high, six feet long, and as wide as the particular vein; hence if the vein was extremely broad, the miner's paycheck suffered. When the fathom policy was introduced into the mines, the checks on Saturday night were, in many cases, drastically cut down. Correspondingly, the grumbling on Telluride's street corners on a Saturday night increased. Often a miner would exceed an eight-hour day under the fathom quota, but his wages would drop below the three dollars per day minimum that was every miner's "pie in the sky" goal.

The mine operators countered the union opposition by importing non-union labor. Disputes, friction, and incidents continued throughout the summer of 1900, each side battling the other one with threats of, and occasionally actual, violence. The British manager of the Smuggler-Union, Arthur L. Collins, as the symbol of management became the target of union hatred. A forceful, aggressive, no-nonsense mine manager, Collins believed in fighting fire with fire. As union aggressiveness increased, so did Collins' stubbornness. Hostilities erupted just before the Fourth of July in 1901. On the morning of July 3, approximately two hundred union miners surrounded the Smuggler-Union mine,[26] and, under the cover of trees and boulders, began firing at the non-union men employed at the Smuggler. The two sides exchanged fire sporadically throughout the day. When the miners returned to their homes that evening, they left behind three men killed and six seriously wounded. The "scabs," or non-union men, outnumbered and beleaguered, surrendered.

Governor Orman of Colorado appointed a commission of inquiry, which after a few brief meetings, decided against proclaiming martial law. The miners and Arthur Collins were induced to initiate bargaining, and after numerous threats, counter-threats, and more haggling, it was agreed that the fathom system would be retained on a voluntary basis. However, in no instance was a miner to work for

[25] Rickard, *Retrospect.*
[26] Wilson Rockwell, "Telluride Story," Telluride *Times,* February 5, 1965.

less than three dollars a day under either the eight-hour day or the fathom system.

After the settlement, the Western Federation of Miners had every reason to celebrate in Fourth of July fashion. Yet the bitterness engendered continued to smolder throughout the autumn and winter of the following year. Incidents of violence occurred, secret meetings were held in the Miners Hall, barroom fights erupted in Telluride's finest emporium, the Sheridan Hotel, and mysterious gunfire rattled in the night.

The tension exploded with the murder of Arthur Collins in November, 1902.[27] Framed by the light from the window of his home, Collins was killed by a charge of buckshot about nine o'clock in the evening, while he was visiting with friends. The union leaders immediately protested their innocence, insisting that the murder had been done by non-union employees in an attempt to discredit the Western Federation. Even some union men darkly whispered that the "venomous" deed had been authorized by other mine owners. In December, 1902, one Harry Jardine, along with other union officials, was indicted for the murder; but since rumor was not accepted as evidence, the court quickly quashed the indictment.[28]

The accusations and counter-incriminations growing out of the Collins murder inflamed what could only charitably be classed as a volatile situation. On the first day of September, 1903, occurred the most serious strike in the entire history of Telluride. Again the union's primary grievance was the fathom system. Union leaders insisted that only the eight-hour day be allowed in the mines and that the fathom operation be abolished entirely. Unsuccessful in his attempt to ignore their demands, Bulkeley Wells shut down the Smuggler-Union soon after the strike began. The editor of the Telluride *Journal*, displaying his knowledge of Latin wrote, "Mining as an industry in Telluride is now *non est*." [29] The *Journal* further editorialized that the mine owners were forced to shut the mines because the cooks in the boarding house quit without notice!

By the first week of November the mine operators, convening nightly and changing the location of their rendezvous frequently, decided that the only alternative was to appeal directly to Governor

27 Telluride Weekly *Journal*, November 27, 1902.

28 *Ibid.*, December 18, 1902.

29 *Ibid.*, September 10, 1903.

Peabody to send the militia. Peabody had the reputation of being anti-union; in historical retrospect this estimation seems accurate. On Tuesday afternoon, November 24, 1903,[30] the first section of six cars pulled into the Telluride depot. As the smart-stepping militia marched through the town, they were greeted with hostile stares and an occasional catcall from the bystanders. The mine managers announced that any men "who..were desirous of going to work" would be furnished "ample protection from violent interference of any nature." Most miners, however, immediately discounted these assurances of "public safety."

The month after the arrival of the militia was marked by violence and arrests. On November 30, thirty-eight men were charged with "vagrancy." Gambling houses and saloons were boarded up.[31] On the surface Telluride was a ghost town, but underneath the facade hatreds were seething. The mine operators continued to appeal to workers with reassurances of protection. However, even they probably realized the ludicrousness of their guarantees against bodily injury. At last they began the deportation of "undesirable" miners, and the importation of "scab" labor on a wholesale basis. Under the cover of darkness on the night of December 8, three poorly ventilated coaches "crowded to the doors" with "scabs" pulled into the Telluride station.[32] With considerable relief, the herded men jumped down from the cars to breathe the frosty night air. Even the fear of harm vanished when they thought of the prospect of spending more hours in the lurching, stale, stench-filled cars. The new arrivals to Telluride's labor force did nothing to alleviate the taut situation. Two weeks after the night visitation of imported labor, groups of miners huddled on Telluride street corners gossiping about the arrest of sixteen Federation members, eleven of whom had been sent to Montrose because of crowded conditions in the Telluride jail.[33] In spite of the tumult many of the mill whistles were blowing by the last day of December.

The new year brought little relief to the beleaguered community. On January 7, the Telluride *Journal* ran the banner "Governor

[30] *Ibid.*, November 26, 1903.

[31] Sherman M. Bell, *Biennial Report of the Adjutant General to the Governor of the State of Colorado, 1903–1904* (Denver: Smith-Brooks, 1904), p. 14.

[32] Telluride Weekly *Journal*, December 10, 1903.

[33] *Ibid.*, December 31, 1903.

Proclaims Martial Law." For the next two months the silence of night was shattered by sporadic gunfire and mysterious explosions. On March 11, on the pretext that all was quiet in San Miguel County, Governor Peabody recalled the Colorado guard.[34]

The militia had not been absent twenty-four hours before a "citizens' alliance" was formed under the leadership of Bulkeley Wells and John Herron, manager of the Tom Boy mine. The alliance, in vigilante manner, searched homes and arrested more than sixty union sympathizers. Exactly a fortnight after they had left, the militia were back in town.

Five days after the guard reappeared, C. H. Moyer, president of the Western Federation of Miners, was arrested as a "military necessity." On April 12, the publicity-minded Adjutant General of Colorado, Sherman Bell, fired off a telegram to his officers in Victor, Colorado: "Take all money on the proposition the *Stars and Stripes* are waving over Fort Telluride and there is no one but Moyer in jail."[35] As a moral warning, intended for the grapevine in, as well as the telegraph out of, Telluride, Bell added, "The people of this county must educate themselves on Treason and read the Articles of War between Sundays, both within and without Telluride Bay and the Harbor of San Juan."[36] Though the officer was referring to the San Juan Mountains instead of the Caribbean, it is obvious that the Spanish-American War had affected his metaphorical sense.

A legal hassle ensued between the attorneys for Bulkeley Wells and C. H. Moyer, finally terminating in the release of all prisoners, including Moyer, on June 15, 1904. On the same day the second expedition of the Colorado guard was withdrawn.[37] As before, however, the withdrawal of the militia did not signify peace. On June 30, Wells, in an open letter to the Telluride *Journal*, indicated that the Smuggler-Union, except for private leases, would close down. The ostensible reason was "the inability of the management to secure a sufficient number of thoroughly competent men to mine the mine fully." Wells laid the blame squarely on the Federation by emphasizing that the union had forced "the entire surplus of competent

[34] Sherman M. Bell, *Biennial Report of the Adjutant General*, p. 15.
[35] *Ibid.*, p. 204.
[36] *Ibid.*
[37] *Ibid.*, p. 217.

miners to find employment outside Colorado," but his motives were undoubtedly more subtle; by refusing to hire Federation members, he hoped to bring the union to terms. If that was his aim, he and the other mine owners were successful. Quiet reigned in the valley for a short period in the summer of 1904 long enough for the owners to secure a full labor force of non-union men.

The overt struggle was finished, but the animosity created left its backwash for a decade after the 1903–1904 strike. The Federation seldom missed an opportunity to stir up labor dissension. The mine owners were just as inflexible in snuffing out the first signs of unrest. In the spring of 1908 passions flared again. By his vigilante anti-union activities, Bulkeley Wells had become the symbol of all that was evil to the miners. On the evening of March 29, Wells was at his club in Telluride enjoying a game of whist with a few friends. Around midnight he left for his home at Pandora, two miles above Telluride where the Smuggler-Union mill was located. He went to bed on the second-floor sleeping porch, and his next conscious moment came about two o'clock in the morning when a bomb exploded under his bed, blowing him to the ceiling.[38] Miraculously, he suffered only cuts and bruises. Perhaps he was all the more grateful for his good fortune when he reflected on the fate of his predecessor, Arthur Collins, several years earlier in the same house. Steve Adams, reputedly a gunman for the Western Federation of Miners, was soon accused as the frustrated murderer by the local populace, but his guilt was never proved. Wells later testified against the leaders of the Federation at the famous Steunenberg trial.[39]

The labor strike in the San Juans was part and parcel of the widespread labor-capital clashes that flared through the mining west in the first two decades of this century. Cripple Creek and Victor were the scenes of the most serious and prolonged strikes in Colorado, the Coeur d'Alenes the most violent in Idaho, and Butte the most fanatic in Montana. Only with the advent of World War I and the public repudiation of the Western Federation of Miners after the murder

[38] Telluride Daily *Journal*, April 2, 1908.

[39] The assassination attempt provoked the editor of the *Engineering and Mining Journal* into a maudlin editorial, which concluded with the ringing challenge "General Wells has known for a long time that he was a marked man, but his personal courage and unfaltering fearlessness have kept him at his post. . . . But how long is this band of assassins to go scot-free?" *Engineering and Mining Journal*, LXX (April 11, 1908), 775.

of Steunenberg, former governor of Idaho, did an uneasy peace settle over the western mining communities.

While the pall of labor warfare hung over Telluride during most of Robert Livermore's first stay, life in the San Juans was far from dreary. E. B. Adams, one of the attorneys for the Smuggler-Union, fondly recalled his days in Telluride during "the glamorous era." [40] Gambling was openly carried on in the saloons: faro, bank poker, and roulette were favorites. The most famous establishment was the National Club, benignly presided over by Morris "Big Mike" Lehman. Adams remembered one gambler who lent a unique touch to the National Club by keeping his pet billy goat, with long whiskers, alongside when he scooped up his cards. Whether or not this mascot was for good luck was left to the imagination of the observer. In sum, Adams remembered Telluride as a "wide open mining town no better, no worse than others."

Congressman Herbert Parson lacked the kaleidoscopic view and the intimacy of Adams' acquaintance; nevertheless, Telluride impressed the young easterner as he took up lodgings one summer evening in the Sheridan Hotel.[41] The saloon in the Sheridan seemed "fair but decorated in somewhat Ichabod style." The natural beauty of the mountains and the chilling air were most refreshing to Parsons. He had made the trip with Fred Farish, a distinguished mining engineer from a family of renowned mining men. Upon arising the first morning after he had registered at the Sheridan, Parsons looked out of his hotel room window at the scene: "a fine view of the head of the valley over some shanties. . . . The high perched houses of Norway are nothing to the airy castles of the mining buildings." That noon the party lunched at the Black Bear mill, where Parsons discreetly watched the miners with approval.

At the mine we had lunch with the men, pretty good looking fellows and all Finns. They work 8 hours, there is no danger, the honest get $3.00 a day, out of which $1. is taken for their board and lodging. They are washed before dinner and the food, soup, peas, potatoes, meat, pie, coffee, etc. was excellent and there was plenty of it. They promote the best men to the mill. A single man can save a lot, but few save anything. They blow all their savings

40 Interview, E. B. Adams, August 13, 1964.
41 Herbert Parson, "Diary," 1910, Parsons Collection, Western History Research Center, University of Wyoming.

getting drunk and generally raising the devil in Telluride. The families of the married men live down there and as living expenses are high in such an inaccessible place as T[elluride], they don't save much. The houses are not pretty, but the women seemed to be well dressed.[42]

Parsons' visit in 1910 came at a time when the labor bitterness was ebbing. One current resident of Telluride, who was living in the valley at that time, thought "a better class of men moved in after the strikes."[43] Parsons' remark on the foreign extraction of the miners was accurate; most of them were Cornish, Finnish, Greek, or Swedish.

All violence in Telluride was not of human derivation: nature also contributed her share to Telluride's brutal catastrophies. Avalanches were an omnipresent danger in the spring. In late April or early May, during eye-squinting, bright blue days, Telluride waited in apprehension for the cannon-like shot, followed by an earth-shaking roar that announced the first slide of the season.[44] Deep scars slashed through the timber covering the mountain sides are mute testimony to the destructive force of past avalanches. Some of the slides, recurring year after year, assumed the names of the veins which they crossed; the more famous ones were the Black Bear, the Liberty Bell, and the Pandora slides.[45] On February 28, 1902, an avalanche struck the buildings of the Liberty Bell mine, crumbling them like match boxes. A rescue party had uncovered a few of the victims when a second avalanche roared down upon them, but incredibly, no one was killed. That evening, as the weary rescuers were threading their way down the mountain, another slide broke loose, killing three men and injuring several others. The church bells in Telluride tolled for seventeen dead in that one unforgettable week.

As with many of nature's disasters, snow slides are marked by freakish incidents. During the huge avalanche of 1903, a rescuer perched atop a tree witnessed a nightmarish scene. As he gazed across the slide area he saw a young woman standing on her porch, frantically waving her apron to attract attention. A slide roared down, the concussion from it sucking her into its path and sweeping

[42] *Ibid.*   [43] Interview, A. R. Gustafson, August 15, 1964.

[44] A vivid description of the avalanche season in the San Juans is in Gertrude Sayre's "Old Smuggler Narratives," MS, Western History Research Center, University of Wyoming.

[45] *Ibid.*

her into vast white unknown. The house was sliced in half, leaving her baby standing in its crib, unharmed.[46] Slides were almost yearly happenings; in 1909 three persons were killed in them.[47] In 1925, Harvey Johnson, who reopened the Black Bear mine, was buried alive under an avalanche and lived. The experience haunted him the rest of his life.[48]

As with labor strikes, Telluride residents learned to acclimate themselves to the interminably prolonged winters of isolation and avalanches. One quickly recognized the most common runways and built his home out of the path and prayed. Life in most mining communities was a compromise, a battle of adjustment, and those who could not or did not find their peace surrendered and left. For those who reached a degree of tranquility by enjoying the Saturday nights on the town, the delights of a sauna bath, the week-long wedding celebrations, the trips into the Alpine countryside, life was more than bearable; it was rich and full. The fortunate remembered the pleasures and forgot the tragedies.

Robert Livermore retained a lifelong feeling that his Telluride years were good ones. When he was a bachelor and succumbed to cabin fever, or after his marriage during the periodic absences from his wife, he could always flee to Colorado Springs, called "Little London" by some, "Newport in the Rockies" by others (both appellations were correct). Here society was high, figuratively and literally. Titled and wealthy British mingled with the *crème de la crème* of eastern society. At a typical party, the guests could easily encompass Count Portales, founder of the now exclusive Broadmoor area and German count; Maurice Kingsley, son of the famous British novelist; and Charles E. Perkins, president of the Chicago, Burlington and Quincy Railroad. The next day, they might be clearing their heads through participation in an improvised polo game at the Broadmoor.[49]

[46] *Ibid.*, p. 12.   [47] Telluride Weekly *Journal*, January 28, 1909.

[48] David Lavender, "Avalanche at Black Bear," *American West*, II (Summer 1965), 32–40.

[49] The best social history of Colorado Springs is Marshall Sprague's *Newport in the Rockies* (Denver: Sage Books, 1961). Other volumes, some helpful and some not, are: James Pourtales, *Lessons Learned from Experience* (Denver: W. H. Kistler, 1955); Francis M. Wolcott, *Heritage of the Years* (New York: Minton, Blach & Co., 1932); Irving Howbert, *Memories of a Lifetime in the Pikes Peak Region* (New York: G. P. Putman's Sons, 1925) and Frank Waters, *Midas of the Rockies* (Denver: Sage Books, 1959).

To any visitor such a society was dazzling, and to a homesick "down-Easter," it was salvation itself. When Livermore first encountered Colorado Springs' society in the early 1900's, the British Empire era had faded. Although Colorado Springs might have lost some of its imperial luster, the economic and social life was far from tarnished. Young Spencer ("Spec") Penrose, of the Penroses of Philadelphia's Maine Line, was a dashing financial buccaneer, soon to be reaping a fortune in Utah copper.[50] Handsome Verner Z. Reed, with his "Scott Fitzgerald" profile, had made his first million from bankers' commissions on the sale of the rich Independence mine to British capitalists. Millions more would soon be flowing into Reed's bank account from the fabulously productive Salt Creek oil field in the center of Wyoming.[51] Personable Alan Arthur, son of the late President, owned the well-designed Trinchera Ranch, which was the playboy's home away from home. Over all this society dominated the patriarchal figure of General William Jackson Palmer, entrepreneurial genius of the Denver–Rio Grande Railroad.[52] As a founder of Colorado Springs, he remained its first citizen even after he was permanently disabled in an accident.

At the General's massive castle-like home, Glen Eyrie, Robert Livermore met his future wife, Gwendolen Young. The Youngs had arrived in Colorado Springs through the patronage of General Palmer, who admired the landscapes of Harvey Young, Gwendolen's father. When Harvey died of tuberculosis, the General assumed responsibility for the children. The General initiated storybook excursions on the Denver–Rio Grande which might last several days. His week-long sojourns in the Sangre de Cristo Mountains, just "picnics," were a particular delight of the Palmers. In the fashion of a medieval baron, he would lead a large entourage of servants, guests, and friends from Glen Eyrie on a bright summer morning, with their destination unknown. To be an invited member of the General's party was the highest badge of social approval one could receive.

Years later, Gwendolen Young's sister-in-law and Jefferson

---

[50] The Penrose family have been described in Nathaniel Burt's *Perennial Philadelphians* (Boston: Little Brown, 1963), pp. 543–549.

[51] Harold Roberts, *Salt Creek* (Denver: W. H. Kistler, 1963).

[52] Robert Athearn has written the latest and most complete history of the Denver–Rio Grande, *Rebel of the Rockies* (New Haven: Yale University Press, 1962).

Davis' granddaughter, Lucy Hayes Young, recalled that wealth did not count so much in Colorado Springs as what one actually made of himself.[53] Like all frontier aphorisms, this probably was a partial exaggeration, but contained a kernel of truth. Colorado Springs' residents did reach the epitome of sophistication by refusing to be snobbish. The worst social affliction one could possess was to be boring. The Springs society valued a bright, stimulating individual, while the formal, the unimaginative, the pedigreed were shunned. Lucy Hayes Young summed up the viewpoint with an epigram she heard as a girl in Colorado Springs: "Colorado Springs was a village full of city people, Denver, a city full of village people."

Whatever the style of society, Robert Livermore immensely enjoyed his interludes in the shadow of Pikes Peak, and with each visit it became more difficult to return to the San Juan country. The Smuggler-Union's leases were not as lucrative, his relationship with Bulkeley Wells was strained, and the feeling that his professional career was stagnating all contributed to his disenchantment with the Telluride. In addition, his family's investments in a variety of mining enterprises in the region were languishing.

Although the properties under the management of Bulkeley Wells were more profitable than many contemporaries assumed, the long strikes and increasing unproductivity of the mines eroded the profits. Because of Wells' reputation, it was just assumed that anything he touched would be financially disastrous. While dividends slowed up, the Livermore investments did return a respectable amount through World War I. Due to the dearth of records, it is impossible to chart the complete history of corporate dividends. However, for the years 1916–1921, the Humboldt mines declared dividends totaling $173,475, on a paid-up capital of $500,000. The First National Bank of Telluride paid returns of $30,000. The Colorado-Superior Mining Company showed an operating profit of $99,616.59 for 1914–1918.[54]

Robert Livermore inscribed *finis* to his journal on several occasions, only to pick up his pen, when the mood struck and the time became available, to write another chapter. Livermore started his autobiography on June 1, 1937, and by December 4, 1950, he

[53] Interview, Lucy Hayes Young, June 27, 1965.
[54] Account book of Bulkeley Wells, courtesy of Thomas Livermore Wells.

had concluded his reminiscences with the termination of his Telluride residence in 1910. After his final retirement, another chapter on Canada was added.

An inveterate diarist for most of his life, he extracted heavily from his lengthy diary entries for his journal. This literary habit not only increases the reliability of his autobiography and consequently its historical value, but also considerably enhances its immediacy and readability. All through the journal there are subtle tones of the "you are there" quality. The original journal and diaries are now in the possession of his son, Robert Livermore, Jr.

In the editing of the manuscript, no substantive changes have been made, only corrections of obvious spelling and typographical errors and occasional changes in punctuation for the sake of clarity.

Historical research inevitably incurs debts beyond repayment. Without the substantial assistance of Robert Livermore, Jr., the editing of his father's journal would have been impossible. His cordial and unstinting helpfulness extended from answering queries and providing more documentation, to entertaining the writer at his comfortable home, Beaver Pond, in Beverly, Massachusetts. At Beaver Pond, Mr. and Mrs. Livermore regaled the writer with family lore of past generations of Livermores; their courtesy far exceeded perfunctory hospitality. Three charming ladies, two in their eighties and one in her nineties—Mrs. Gwendolen Young Livermore (Mrs. Robert Livermore, Sr.), Grace Livermore Wells (Mrs. Bulkeley Wells), and Lucy Hayes Young (Mrs. George Young)—all were unfailingly generous in their personal reconstruction of events of a half century ago. Thomas Livermore Wells, by allowing access to his father's account books, clarified some of the economic struggle of the last days of the Smuggler. He also graciously took time from his busy schedule to read this Introduction.

Since the Smuggler-Union and Telluride were and are synonymous, many residents of the Telluride-Ouray area (referred to with pride by the local citiznery as "the Switzerland of America,") recalled much of the local color of the Livermore era. E. B. Adams, one of the Smuggler-Union's attorneys, whose successful suit against the neighboring Liberty Bell mine for $403,853.08 is still the talk of Telluride, has a Dickens-like gift for vivid vignettes and character description. The writer had the privilege of sitting in "E. B.'s" living room on three separate and lengthy visits, listening to his penetrating analysis of mining history.

In the town of Telluride, Silvo Oberto, whose family have long been associated with mining in the San Juans, acted as host on one brilliant Saturday. A. R. Gustafson, present mayor of Telluride and long-time power company official, gave invaluable data on Telluride's economic fluctuations. Mrs. George Wagner, whose husband was one of San Miguel County's most successful mining magnates, answered questions on a busy day. Homer Reid, of the Busy Bee Corner Drugstore, furnished copies from his rich cache of photographs.

In Ouray, A. C. Hilander, general manager of the Idarado Mining Company and present-day owner of the Smuggler, spent several evenings describing the contemporary background of mining in central Colorado. Fran and Keith Johnson, whose personal history was closely identified for so many years with the Camp Bird mine, guided the writer with their comments through the labyrinth of local mining lore. Two Ouray attorneys, Jerome Paul and Philip Icke, clarified the highlights of several famous litigations. Mrs. Rosa Zanett related the activities of her husband, long prominent in the mining industry of Ouray county. Misses Alpha and Hester Sigfrid provided documents and reminiscences pertaining to their father's life: Carl J. Sigfrid was an early and very successful Western Slope attorney.

In Montrose, Harrison Loesch, Dan Hughes, and William J. Knous, members of the bar of the Western Slope of Colorado, offered their reminiscences and suggested several very valuable contacts.

Three sons, whose fathers were all mining engineers in the Ouray-Telluride vicinity sometime during their careers—Charles N. Bell, Jr.; Robert Sayre, Jr.; and William Spencer Hutchinson, Jr.—all presented manuscripts and reminiscences of their families' association with early Colorado mining. Mrs. Harold Worcester donated the significant correspondence between her husband and Robert Livermore, letters which graphically outlined the liquidation of the Livermore holdings. John Parsons, Manhattan attorney and historian, transcribed the diary of his father, Congressman Herbert Parsons, recounting his early twentieth-century journey to the Rockies. Mrs. Joyce Dion uncomplainingly typed the manuscript several times.

Finally, to my wife, Joyce and our two siblings, Deborah Ellyn and David Randolph, my gratitude for putting up with a historian in the house.

GENE M. GRESSLEY

I

Boston Beginnings

I have had a lot of fun out of life. I have often thought that someday I would like to put my experiences down in writing, but autobiographies are for people who have taken their part in affairs of public interest, while I have been more or less on the fringes of things. Then, too, the thought of writing for publication dams the current of my narrative, and in times past when I have tried to set down what I have done and seen in orderly enough form for possible readers, the result has not been encouraging. Just the same, the urge keeps recurring, and I have promised myself that when I have settled down at my farm in Boxford (retired is the word) I will try again to put on record, if only for my children's reading, the story of a life if not distinguished, certainly a varied one, and covering a period of change in this country which is already beginning to take on a historical tinge. I used to think I was born about thirty years too late, as otherwise I would have been a sailor when sail was still supreme, a plainsman when there was a real frontier, and a miner when new goldfields were in the making. Nevertheless, my imagination would not allow that the glory of those days had all departed, and one after another I tried them all, to find that though the palmy days were over, plenty of the flavor remained.

The nineties saw the end of an era at sea and of the frontier. Steam had nearly banished the sailing ship, and dry farming, the wire fence, and final stroke, the auto, changed the last strongholds of the carefree boisterous West that I first knew to the standardized highly modern land of today. I am glad that I saw even the tail end of those eras, and like Caesar can say "Quorum pars parva fui."

Now that I am sixty, and may be a lot older before I finish this, I

1

feel that it is about time to get started, as retirement seems as far away as ever, and I am beginning the yarn not in the quiet of the "Farm" but on the road, in pursuit of my profession of mining. Fortunately, I always had the habit of keeping a diary, which with many gaps provides the framework on which I can build.

My people, the Livermores, are good stock, descended from the original immigrant, John, who settled near Boston in 1640. Our branch early emigrated to New Hampshire, and in Wilton and Milford were for generations respected and prominent citizens in their communities, mostly in the professions as doctors, lawyers and ministers. My father, Thomas Leonard, started the family peregrinations early in life, crossing the plains to California with his family in 1849 at the tender age of five. Their stay was brief, and 1861 saw him enlisted as a private in a New Hampshire regiment for four years of war, from which he emerged as colonel. When I think of his life, so full and well rounded, and compare it with my own incompletions, I feel humble enough to leave my song unsung; but as that is not the spirit in which to take up this record, I will leave his achievements to the memory of his family and friends, and keep to my personal story.

After the Civil War, my parents settled in Boston, and there we children, four of us, were born, there my mother died when I was three. We then moved to Manchester, New Hampshire, where for five years my father was agent of the Amoskeag Mills. My childhood life there and in Shirley Hill where we spent our summers, simple New England farm life at its best, has given me a lasting love of the country and now of this part of it in particular, to such effect that no matter where I lived and worked afterward, a country home hereabouts was always my goal. Having owned this place in Boxford for some nineteen years, I begin to think that goal at least is achieved.

When I was eight we returned to Boston and lived in Jamaica Plain, where the roomy, comfortable house my father built was home for many years. We lived in the pleasant fashion of suburban Boston, with horses to drive and ride, summers at the shore, and ample friends to play with. Our place was the mecca for "the fellers," my gang, whose mischief sometimes brought long-suffering wrath on my head. Father was the soul of hospitality and loved to entertain not only his legion of friends but the friends of his children, whom we brought home in increasing numbers through school and college.

From this house we made our many excursions, and in it we always found haven, until at length, when we had all married and set up our own establishments, he found it too lonely, and moved in town.

Father was a great believer in the public schools and sent me to the Boston Latin, while most of my friends went to private or semi-private schools. Consequently, my friendships being pretty well fixed, I did not make many friends at public school, and was not particularly happy there. I did get a thorough grounding in the essentials of Latin, history and the three R's which has stood me in good stead, and for that I am grateful. However, I did not do well in classes, partly no doubt due to an invincible dislike of home lessons, but partly to interest in subjects outside the curriculum such as writing stories and drawing. Often when I should be studying in class, behind my desk I would be sketching in pen and ink some imagined adventure, and inevitably got caught and chided. Perhaps if there had been more of the modern idea of encouraging a pupil in his bent instead of holding him to a strict routine, my small talents might have amounted to something; but that is guesswork.

The wanderlust took hold of me early. I wanted to go to sea. Why, I don't know, as except for an uncle in the navy, and a stray ancestor or two in the old naval wars, ours was not the sea tradition. I guess Marryat and Dana, my favorite authors, had much to do with it.[1] Although in 1892 America's glory in sailing ships had long departed, and the day of sail itself was waning, to my mind the romance of Cape Horners and tall packet ships still existed, and nothing would do but for me to ship before the mast in a square-rigger. I spent all my spare time and some that I shouldn't have spared hanging around the docks at Atlantic Avenue, boarding ships, hobnobbing with sailors in their forecastles, trying my hand at climbing rigging under the lenient eye of the mate, and breathing the incense of tar and cargo. Lessons were a burden, schoolboy sports were stale; I wanted to be up and away.

In those days Atlantic Avenue was a far different place than now, when you can hardly see the water for elevated ways and housed-in piers. Sailing ships, a good proportion of them square-riggers, still tied up at India and Constitution wharves; T wharf still swarmed

---

[1] A prolific novelist, Frederick Marryat (1792–1848) wrote about the sailor's life. Richard Henry Dana (1815–1882) is best known for his *Two Years Before the Mast*.

with fishermen, Gloucester schooners, trim and tall-sparred; and
sailormen, unmistakable by their gait and dress, wandered along the
Avenue. There was plenty of the salty flavor to excite a youngster's
fancy.

I fear I made my father's life miserable with my begging to be
allowed to go. I must have seemed very young. I was only fifteen,
and none too husky; but he had the recollection that he had enlisted
for the war at little older, and wise man that he was, knew the best
way to get rid of an obsession was to try it out, so gave his consent
when and if a suitable ship came along. Knowing what I later learned
of forecastles, perhaps it was lucky for me that about that time the
Massachusetts Nautical Training School had come into existence,
with the purpose of training young men with a seafaring leaning to
become officers of our hoped-for merchant marine.

The navy had lent a wooden frigate, the *Enterprise*, built in the
seventies, and staffed with a full complement of commissioned
navy officers, with the stipulation that the boys should be given a
naval training and should serve as a reserve for officers of the navy in
case of war. The ship was being refitted at the Charlestown Navy
Yard as a "jackass frigate" (why the uncomplimentary name, I
never knew) with flush decks, that is, instead of the former waist
broken by poop and forecastle. She carried a full barque rig, fore
and main square, and mizzen fore and aft rigged, and had auxiliary
steam with a smokestack which telescoped when under sail. She had a
battery of old-style muzzle-loaders, Parrotts I believe they were
called, served from the gun deck through ports piercing her sides.
To complete the description, she was about 375 feet long, and 1375
tons burden.

With some reluctance I gave up my desire for the forecastle, no
doubt somewhat influenced by the blue uniform in prospect, took
the simple examination required, and was duly enrolled with some
ninety others as a cadet in the service of the Commonwealth.

It was early in April of '93 that I went aboard. I remember there
had been a light snow, and I looked aloft at the slippery rigging and
wondered if I would ever have courage to climb the shrouds. I was
given a hammock, stowed in the hammock rail by day and slung
on the berth deck at night, which as soon as I got used to lying on my
back in a curved position made as comfortable a berth as I could
want. Although it was the first time that I had been away from home,

I don't think I was homesick at all. The routine of shipboard life which began at once, the process of making new acquaintances, the novelty, and the fulfillment of a longing, all kept me free of regrets; and when after a week or so we were given a brief shore leave, I swaggered around at home like a many voyaged mariner, though I had as yet to leave port, and after satisfying my ego among admiring stay-at-homes, left for the ship with satisfaction.

The *Enterprise* was a tidy ship. Spick and span from her new overhaul, white-painted, sails new from the loft, rigging taut and freshly tarred, she was a handsome sight for a sailorman, whether under steam with sails furled on the squared yards, or under sail with all canvas set and drawing. After all these years, the names of some of her officers and most of the cadets escape me, but I shan't forget Commander Merry, afterwards an admiral, firm yet understanding captain of a shipful of sometimes unruly youngsters; Lieutenants Forster and Osborne, aloof navigator and stern executive; Ensign Miller, young, with a twinkle in his eye, whose special charge we were; mate Tyrell, rubicund seadog, with his "Rise and shine, tumble out now" at ungodly hours; Nelsen, Norse second mate, who drew correct pictures of full-rigged ships in my diary; the Andersons and Ericksens of the paid crew; and later, Jack Greenwood, trim British man-o-warsman, shipped in England, who made himself my special mentor in the arts of marlinspike seamanship.

My fellow midshipmen were a good average lot of American boys, drawn from all over the State, not, as one would expect, mostly from seacoast towns. Some were sons of seafaring men, intending to follow their fathers' trade, but most, like me, were probably inspired with a love of change and adventure, which this first cruise of the training ship seemed to offer. I was with few exceptions the youngest of the lot. Few or none of us had any sea training, and for handling ship a foremast crew of able seamen had been signed on. As we became more expert these men were gradually laid off, and by the time we sailed across the Atlantic only a few were kept for leaven. From then on, we "handed, reefed and steered" as well as any A.B.

The roster of the cadets' names would probably read far differently now, in days when the Crzegelskis and the Magaluppis are as good Americans as any. My diary shows autographs of Bennett, Lincoln, Manley, Jordan, Whitney, and the like, with a sprinkling of O'Leary, Moran and Mock for good measure. Although I have hardly seen any of them since, their names bring back their appearance and qualities to mind, and I remember with pleasure for the most part my long contact with them. Folger, son of a Boston pilot, the best seaman of my guncrew, a mighty man on a yard arm; Cushing the bugler, who always missed the same note in the same call; Lawton, looking for romance like myself, afterwards a captain in the Philippines; Reed, who could reel out a string of oaths of innocent intent to the admiration of all; Foster, as young as I, with whom I was matched for many a boxing bout in the dog watch; Maguire, of the Worcester Irish, and a good friend; and many another, some fine, some colorless, and a few, poor stuff, but on the whole a good company.

The whole ship's company was divided into port and starboard watches, and the cadets were divided into crews of ten each to whom was assigned a gun and a boat. Mine was the second crew of the port watch, with the large sailing launch as our charge. The usual healthy rivalry developed between these crews, and each believed itself to be the best on board. I had the proud position of captain of my crew, though being too young and far from a leader, I don't know why, unless my enthusiasm for everything to do with sail, and a certain agility aloft later acquired had something to do with it.

In port only a day watch was kept, but at sea the usual four hours on, four off was the rule, with the two dog watches of two hours each to change the order. I think one of my most poignant memories is the misery of being called at four A.M. to stand watch, when all my youthful being yearned for sleep. Our uniform for dress and shore leave was the navy blue tight-fitting braided jacket of the period, the ugliest stiff visored caps ever designed, and a pea jacket for cold weather. For sea duty, it was the loose-fitting, comfortable white ducks of the seamen, into which we eased with a sigh of relief when off parade, and often in informal ports managed to wear ashore.

Drills, first lessons in deck seamanship, and when we had made ourselves used to heights, sail drill, took up our first days in port. How well I remember the first cautious climb up the fore rigging, through the lubber's hole into the top, and then, spurred by sarcastic yells from the mate on deck, "Now don't squeeze the tar out of that riggin'," edged out on the foot ropes to the end of the topsail yard. It wasn't long before we were running up the ratlines at the cry of "Lay aloft," and actually standing on the yard to furl sail.

Finally, we cast off for our first practice cruise, down harbor and into a choppy sea off Boston light. By the time we reached Province-town, our destination, we landlubbers had turned ourselves inside out and lay about exhausted on deck, little cheered by mate Tyrell's remark, "Now that ye've got all the shore swill out of you, you can fill up on lob scouse and be good sailormen." I, for one, looked with loathing on the mastheads wildly whirling against the sky, little believing that a few weeks later I would be sitting on the royal yard enjoying the giant swinging of the foremast in the Atlantic swell.

The first month or two was spent along the New England coast, handling sail, boat and gun drill, making port under steam and sail, and learning fast to be sailormen. Some incidents stand out, as when off Gloucester in a rain and sleet squall we swarmed aloft to reef topsails, and I, jumping from the shrouds to the footropes, lost my hold and fell, by luck hitting the lift just below, and hanging on for dear life, only to continue my climb, too sheepish in the face of jocular remarks to stay scared; another, when one balmy day we lay at anchor in Gardiner's Bay, and all boats were called away to seine for fish. We made a great lark of it, lowering boats in smart style, stretching nets between each pair, then racing to shore to haul our catch out on the beach. The take was small, but the luxury of a naked swim and a sun bath on the lonely beach was well worth the trouble. I recall New Bedford of that day, its whaling days scarcely over, with numbers of old whalers tied up and rotting at the wharves, one lone one still fitting out for a cruise, and a tiny bark from the Azores, jammed with colorful brown humanity, with, for a home touch, a cow, pigs and chickens installed on deck. We explored those old ghost ships of the past, climbing at some risk their rotting rigging, and wondering at the primitive steering gear of wheel ropes running from a visible tiller, the drying-out ovens, and the round forecastles built on deck. Except for the sole survivor, stuffed and mounted as it were, set in cement on a rock ashore, not one is now left.

There was to be a naval review in New York that year, a belated celebration of the landing of Columbus, to which the fleets of other nations had been invited, and much to our joy we learned that we were to be there. This was to be a sort of farewell flourish before sailing for the other side, and our officers no doubt thought we were seasoned enough salts not to do them discredit, which indeed proved to be the case. Most of us I suppose had never been away from New England, and to me the thrill of seeing New York, and under such auspices, was enough to keep me awake in my hammock many an hour when I ought to have been asleep.

We came through Long Island Sound, and on approaching Hell Gate lowered top gallant masts to pass under Brooklyn Bridge, in which process, I as one of the foretopmen was aloft when we went under. Youthful enthusiast that I was, I shinnied up the bare pole to the masthead and, hanging onto the truck, waved my cap at the grinning faces on the bridge. I wish I could do that now. Rounding the Battery, we took position upriver above the two lines of men-o-war anchored there, ships of many countries, seeming to us tremendous engines of war, what today would look like antiquated gunboats. Most of them depended to some extent on sail, and some were more or less full-rigged for sailing.

Our navy then consisted of a few relics dating back as far as the Civil War, and the newly built "White Squadron," so called, of which the nation was immensely proud. These were our first attempt at modernizing our navy, a small enough beginning for that long-neglected arm of the service. The flagship was the *Chicago*, a steel cruiser, ram-bowed, fitted with stumpy bark-rigged masts. The *Atlanta, Boston, Yorktown,* and *Bennington* were smaller, down to gunboat size, and all rigged with some brand of auxiliary sail. Years later, I saw "Teddy" Roosevelt's "Around the world fleet" end its long journey in Hampton Roads, and thought with amusement of the fleet that seemed so grand in '93; and now if I could see the Roosevelt fleet they would look as antiquated as the old one.

We were really not part of the review, hence did not participate in the various ceremonies of the fleet nor in the grand parade on shore, but were given shore liberty, when we explored as much of New York as we could in an afternoon, (the Eden Musée being the mecca for most of us) and had a naval parade of our own when we weighed anchor and, manning the rail at attention, with squared yards, steamed proudly down the lane of ships and received their salutes, then breaking out sail when still in their sight, headed for home in a cloud of canvas. I have an idea our officers were quite tickled, judging from their subsequent good humor, by the smart way at the command "Lay aloft" we ran up the rigging fifty strong, loosed sail, and sheeted home.

In Boston we overhauled ship and laid in stores, and on July first steamed down the harbor outward bound. Once clear of the land, we broke out all sail, not to be furled again till the "Lizard" was sighted. I still get a kick, old traveller that I am, at seeing new places, but it is nothing compared to the vivid anticipation of a youngster of sixteen on his first trip to foreign parts. Even at the start everything was novel and interesting. The land-free horizon, the deep-sea swell, a whale passing so near that we could see his wicked little eye and the barnacles on his back, Mother Carey's chickens, which took the place of the shore-loving gulls, the occasional sail, and above all the thought that the next land we would see would be something strange and different. In a passenger ship of today the voyage is over almost before it is begun, and one is so housed-in and far from the water that nothing much can be seen anyway, but a sailing ship slips leisurely along making no commotion, and marine life gathers around.

There was no idleness for us at sea. True, drills with rifles and formation exercises were dropped, but the infinite variety of sea jobs took their place. Scrubbing decks in the morning, holystoning them for good measure every few days, making chafing gear, tarring down and setting up rigging, sending down sails and bending new ones, working the ship, with all the watch aloft or at the braces, lessons in navigation, when we learned to shoot the sun and work out our position, all kept us out of mischief.

Still, there was time for enjoyment. It was fun to race barefoot up the ratlines when a job of reefing was to be done, and see who would be first out on the yard, where the winner would "jockey the yard arm" beyond friendly rope holds and sing out to his slower mates "Haul out to leeward!" and on a warm, pleasant day for a couple of us sent aloft to the royal yard, after finishing the job, to sit there, one arm around the slender mast, "yarning," and looking down on the bellying canvas beneath, the white decks patterning the vessel's lines, the sharp bow breaking foam in the blue gulf stream waters, until a keen eye on the quarterdeck would spot us, and the call "Lay down from the royal yard there!" brought us to deck and duty.

The thing I liked best was steering. I took to the helmsman's job naturally and looked forward to my trick at the wheel. There was a feeling of power and accomplishment in keeping that bulk of wood and canvas to her course, by day to see every sail drawing cleanly when the course was "full and bye," and by night when the sails were dark masses blotting out the stars to watch the compass needle swing true in the binnacle light, and feel the ship respond to the turn of the wheel.

The Atlantic, even in summer, didn't spare us discomfort entirely. Graphic extracts from my diary say for July 4th, "At noon, it clouded up and a big rain squall passed over us soaking everyone. All hands were called to furl sail, and I went up the foretopgallant. I had a mighty hard time till Ericksen came up. At six o'clock the spanker jibed and brought her up into the wind which took every-thing aback, but we braced around and soon filled away again." And on the sixth, "At one bell in the evening the rain commenced to pour down and the lightning flashed all the time. I was sent up the royal yard to furl sail. Then we took in topgallants. About twelve, the squall burst on us and the lightning was incessant. The water shone with phosphorescence and the rain came down by the bucketful. When we turned in, I tell you I was glad to get off my wet clothes." And again, "Wet and foggy. At seven bells both wheel ropes parted and we luffed up, but as there was no wind to speak of, no harm was done. It was raining pitchforks and we had to jump aloft and furl the mainsail. It was soaking wet and stiff as a board. We spent an hour of the hardest work I have ever done, but finally got the bunt up by means of the top burton and passed the gaskets. My hands were so parboiled I could hardly hold on." Old-timers who remember the weight and the iron-hard consistency of a wet main course will appreciate the job we had.

The dog watch at eight bells in the afternoon was a time for "skylarking." Generally, boxing matches were arranged, at which I was a frequent contestant in the bantam class. Sometimes a har-monica provided music for a hornpipe by one of the crew. Once or twice there was a grudge fight, which, if fairly fought, the officers, usually interested spectators of the other sport, would discreetly fail to see, believing that bad blood would best be worked out.

A sail was always of interest. The lookout's cry of "Sail on the port bow" brought all idlers to the side to speculate on her rig and nationality. We were off the passenger steamer lanes, and the usual passer-by was a freighter. Many were sailing ships, barks, brigantines, and a full-rigged brig or two, a rig even then nearly extinct. All the steamers had sail of sorts. As we neared the English Channel the traffic grew thicker. There was a fresh northerly breeze blowing, head wind for us, but fair from the North Sea ports, and ship after ship appeared over the horizon, big four- and five-masters, many of them. First a gleam of white on the sea rim, then hull up, then foaming by us, a "bone in her teeth," towering pillars of canvas, every rope taut and humming, outward bound.

When at last we sighted the Lizard and the green shores of England, though the excitement of seeing a new land was there, somehow it seemed curiously more like coming home after a long absence than to a foreign shore. Probably my ancestral blood and familiarity with English books had something to do with it, but I never had then or afterward any feeling of foreignness in England. It is long since I have been there. Perhaps I would, now.

We coasted along the chalk cliffs with their green-topped slopes all day, passed the Needles with their striped lighthouse and cast anchor in Southampton Harbour just twenty-five days out from Boston. My diary briefly records for the following day, "Scrubbed paintwork all day."

We were given five days liberty, and after drawing our meagre shore money, three of us toured as much of that part of England as time and expense allowed us. We went, like a busman's holiday, to Portsmouth, to see the navy yard, where we struck up acquaintance with some English tars and toured the music halls, then to London, where we arrived at three A.M. and bunked on the seats of the station until the town woke up. We saw most of London streets afoot, boylike asking our way of everyone, and everywhere meeting kindly interest, and wound up with a visit to Westminster, of which I dutifully set down a description which reads now as if it was cribbed from Baedeker. Then back I went to Portsmouth, where my sailor friends made me welcome and took me to a neighborhood party made up of their best girls' families and neighbors, brought in to meet the "Yankee midshipman," a very respectable gathering of nautical flavor, at which after refreshments everyone rendered some ballad of the sea.

Liberty over, and money spent, the *Enterprise* received us for a few days of routine, mostly scrubbing decks already immaculate, navigation and boat practice all over Southampton harbour and part of the Solent. The *Chicago*, pride of the navy arrived, and anchored nearby, and some of us went over her. I remember that most of her crew were Scandinavians, native Americans not taking to the sea much at that time or for a long time after.

Lisbon was our new port. The Bay of Biscay according to custom gave us a rough welcome. My remembrance of that trip is a succession of cloudy skies and choppy seas, punctuated by an episode of being sent out on the jib boom to furl the flying jib and being hurled first skyward, then dropped onto the spray of an icy comber, but managing to get the sail furled after a fashion.

After nine days of beating against head winds, we came to anchor up the Tagus River[2] near the *Bennington*, which had preceded us, opposite the city rising to its citadel-crowned hill, promptly to be surrounded by bumboats all shouting their wares. My British man-o-warsman friend, Jack Greenwood, who had shipped in Southampton, and I, toured the city on our Sunday liberty, and no doubt Jack, man of many ports, kept me from being imposed on by the shore wolves who abounded in this really foreign land, far different than homelike England.

I mention in my "log" the fine buildings and large squares with their sculptures and monuments, street scenes strange to me, of men with gaily-colored sashes and broad-brimmed hats, pretty girls with baskets on their heads, and donkeys buried in their loads. We climbed to the citadel, where I described the view as "Away to the north lies the most thickly populated part of Lisbon bounded by mountains on one side and the Tagus on the other. The surrounding country is dotted with farm houses very dazzling in the sun, as they are all whitewashed. Far out in the harbor was a solitary little steamer with its wake showing like milk astern. Farther down, the *Enterprise* and *Bennington* lay, surrounded by a mass of small craft. In all directions lateen-rigged fishing boats were drifting in from their day's fishing, their sails idly slatting and their nets stretched against their sides."

There was a night bull fight which we attended, and I described it at great and dramatic length as an event of importance, but I omit my somewhat florid description, as it was like all bull fights except that in Portugal they didn't kill the bull, but overwhelmed him with padded clowns and ushered him ignominiously from the ring with oxen.

2 The Tagus is the longest river in Spain and Portugal. Navigable for most of its 625 miles, it empties into the Atlantic Ocean near Lisbon.

Anchor weighed, catted and fished, and down the Tagus, five days' drifting brought us to Funchal, Madeira,[3] where we gazed with delight at the emerald-green hills rising abruptly from the sea. No harbor here, but an open roadstead, where lay at anchor the *Saratoga*, full-rigged ship of the Pennsylvania Nautical School. Her boys were in swimming overside, and hardly had we cast anchor when the boat booms were run out, lifeboat launched, and we too were diving from the channels and rails and splashing about in the warm, clear water to our hearts content.

Here we saw lying inshore at anchor the *Red Jacket*, so I was told, last of the old "Black Ball" line of New York–Liverpool sailing packets, whose chantey, the "Black Baller," I sang many a time at sea, and do yet for a dinner amusement.

Given shore liberty at last, we joined the *Saratoga* boys ashore, and with them thronged the town, riding in the canopied ox-drawn sleds over slippery cobbles and, sailor-like, hiring horses for a ride up the steep mountain streets, a guide hanging onto each horse's tail; then boarded the sleds for a coast down, steered by side ropes in the hands of running men. No doubt tourists do the same things now.

The outstanding event of our stay in Madeira was the battle staged by the combined *Saratoga* and *Enterprise* crews and the populace. How or where it started, I never knew. The first I knew of it, I was sitting with one Charlie Stoll, a new-found *Saratoga* friend, in a *posada* on a little plaza, when the sound of angry voices and running feet brought us out to find a milling crowd surrounding a few sailor lads. Canes and fists were flourishing, a knife or two glistened, and things looked bad for the embattled youngsters, but down every alley and street came sailor reinforcements, enthusiastic for whatever fray was brewing. The tide of battle soon turned and a solid front of blue jackets, my own and Charlie's in the forefront, was pushing the citizenry back across the square, our skilled fists too much for the ineffectual blows of the natives; when suddenly the local *gendarmerie* appeared on the scene, and we found ourselves confronted by a businesslike array of uniforms and drawn sabres. The clash was brief, a blow or two was struck, then discretion prevailed, and the American forces melted (again Charlie and myself in the forefront), leaving a few prisoners in the hands of the enemy.

[3] Funchal, located on the southeast coast, is the chief port of the island of Madeira.

I believe consular and officer intervention, by what arguments I don't know, persuaded the authorities that it was nothing but youthful ebullition of spirits, and the prisoners were released in the day, to be escorted in triumph through the streets by their massed shipmates. That night aboard, some of us nursing honorable bruises, we told the story over and over and boasted of what we would do to the "Portugees" when we got ashore again. On the whole, looking backward, I guess for peace-breakers we were treated very leniently.

Next day a square-rigged man-o-war was seen becalmed and in tow by her boats. She was the *Portsmouth*, U.S. Navy apprentice training ship. We lowered our boats and raced joyously to the rescue, only to be beaten to the goal by a tug which took her in tow. We made fast astern and amused ourselves jumping overboard in tow of a rope trailed, disregarding possible sharks. We were given shore liberty once more, and, the *Saratoga* boys being quarantined because of their part in the row, we chummed up with the *Portsmouth* lads, and did the town again. Though our money was as welcome as ever, many hostile looks reminded us of the clash of yesterday, but there was no action until at night. We had gathered on the pier to wait for the ships' boats, when a shower of cobblestones rattled around us. An enthusiastic rally was made, in which I believe I was among the leaders, and at our charge the dark mass of assailants vanished up the seafront byways, leaving us with a barren victory. I remarked, "Everybody was glad to get up anchor and sail for the Canaries next morning, for the town of Funchal didn't seem very hospitable to us; only to our money."

Four days from Funchal we cast anchor in the harbor of Las Palmas, Grand Canary, where the hot winds from the desert coast of Africa not many leagues away made us thankful to don clean white ducks for shore leave and to discard even those as soon as we could find a secluded beach for a swim. I remember little of the place except that the town was dull, and we were nearly penniless, so could not avail ourselves of its few amusements. One episode stands out. There was a little fort hard by the beach, constructed of rough stones, locked and deserted. Curiosity, and the tempting footholds offered by the jutting stones of the sea wall, made me volunteer to scale it and open it up from inside. The going was easy for the first thirty feet, but the stones became smaller and slipperier, and ten feet from the top I found myself unable to go forward or back. By that time my hands had lost their grip, and I was in something of a panic. My shipmates saw my plight and rushed for help, but before they could find it, a final desperate effort put me in grasping distance of the battlement, and with my last strength I drew myself up and lay panting on it. I found my way down the staircase and out some way, vowing not to try that kind of fun again.

Early in September we heaved up anchor and set sail for home. It was to be thirty-one days before we were to sight land again, and in that time we polished up our growing knowledge of seamanship with every variety of handling ship. We were soon in the trade winds, when day after day went by with steady fresh winds from aft. Topgallant and royal staysails, hitherto in the locker, were broken out and sent aloft, and makeshift studding sails, made from old jibs, were set. These were the bane of our existence, always having to be taken in and set again, or else carrying away their sheets or halliards, when all the watch would be hard put to it to get them inboard and reduce them to tameness. I can hear them thundering, and the call, "Lay aft to the braces, stand by to take in stu'nsails!" yet.

I have often wondered in what lies the charm of a book like Dana's classic *Two Years Before the Mast*. Once when a sailor friend tried to read it he said, "I don't see what there is in this book. It is just what happens every day." I guess that is the answer. There was certainly no attempt on the author's part at artistic writing. He told just a plain tale of events familiar enough to a large class of the population of that day, but its very lack of attempt at sensation brought conviction of its truth and gave a picture right in every line, both to those who knew the life and those who lived so differently that it stirred their sense of real adventure.

As I read over my "log," immature as my descriptive powers were, to me also it brings back the qualities of life at sea on a sailing ship, and the deep impression it made on me as a boy of seventeen. True, we had no extraordinary adventures. We weren't subject to the whims of brutal ship's officers. Our food was monotonous but wholesome. Beyond a squall or a half gale or two, we had no weather to fight. But there was a charm to the life nevertheless which appeals to something deep in the blood.

One thing my search for romance at sea did not allow for: its monotony. Day after day passed by sailless; the same blue skies by day, the same gathering of the picture clouds of the trade latitudes at dusk; the creaks and straining of the hull and rigging; the crash of the lee bow wave as the sharp stem lifted and fell; the same white-capped waves as far as the vacant horizon. It was then I thought of home and all it meant, and vowed that once there, I was through wandering. In such circumstances the smallest things took on interest. A school of bonita one day appeared under our bow, chasing flying fish. An old hand armed himself with a pair of oilskin pants for a game bag, and a line and hook with a rag attached for bait, and promptly hooked one, the only one. A black dot was sighted astern, and we put about, thinking it might be a ship's boat adrift, only to find it a barnacle-covered piece of wreckage, two guardian sharks swimming below it. A school of grampus, bulky and leisurely, crossed our bow, undisturbed by our slow progress. A rare sail came up from astern near enough to make her out a full-rigged ship, skysails and stu'n sails set, and left us as if we were anchored.

Routine and the numberless jobs about the ship was the cure for homesickness and grumbling. By daylight we were scrubbing decks; then breakfast of porridge, molasses, bacon, hardtack and coffee. The morning watch went about various work, setting up slack rigging, replacing worn ratlines, tarring down standing rigging, making punch mats and other chafing gear, polishing bright work, and on calm days painting ship oversides slung in a boatswains chair. The old proverb "Six days shalt thou work as hard as thou art able, and on the seventh holystone[4] the decks and scrape the cable" was fairly close to the truth. Occasionally the shrill squeal of the boatswain's pipe summoned the watch to tail onto the braces, or rarely, that important maneuver of tacking ship, when every man had his station, and at the commands "Up helm," "Loose tacks and sheets, haul mainsail" brought her about as smartly as a crack yacht crew.

After dinner, usually salt meat or stew, rice, some resuscitated dried fruit, and the inevitable hardtack, the watch below often had navigation to study. Although I learned to take sights and work out our position, mathematics was never my strong point, and I made pretty heavy weather of it. Once a month we were given our rating,

---

[4] A soft sandstone, holystone was utilized to scrub down a ship's deck—not exactly the favorite task of most seamen.

with a maximum of five. Mine for August were: Seamanship, 5;
Steering, 5; Reel and Log, 5; Navigation, 3.5; Conduct, 4; which is a
fair sample of how I stood in the estimation of my superiors.

Without making too many quotations from my Log, for local
color an occasional extract may add to the picture, of which the
following are samples:

September 16th. Starboard watch holystoned the gun deck.
(This, with glee we being in the port watch as it was a hated job.)
It looks like a blow, and we are pitching and rolling a good deal.
This afternoon we had a bad time with the stu'n sail. The jumper
carried away, and in a moment the boom was in mid air almost
to the foreyard. In hawling [*sic*] in the sail, Anderson the seaman
caught his hand in the sheet and had it lacerated badly. He went
below and we finally got the sail in, after towing it alongside for
about half an hour. September 26th. A fine fresh breeze and
a pleasant day. I was sent up to ride down the mainstay and give
it a coating of tar. It was rather ticklish work, as the vessel had a
noticeable roll on and I had no boatswains chair. However, I
managed to do it without leaving a single "holiday," and came
down. The 27th. Squally and unpleasant. I was at work rattling
down the mizzen all morning. The 28th. At four in the morning
when we turned out, there was a high wind and a dark stormy
looking sky. The starboard watch had been having a hard time of
it. They had to furl royals and topgallants'ls and let go the upper
topsail halliards, bringing the vessel down to single topsails and
fore course, the smallest sail we have yet had. The port watch had
their coffee standing in groups around deck and shivering in
their wet oilskins. The 29th. In the midwatch last night we were
waked up by the word "Come rouse out here, port watch furl
the foresail." We slowly climbed the rigging, trying to wink the
sleep out of our eyes and furled away. After that, the word was
passed "Furl all sail before you lay down"; so I went up to the
to'gallants'l[5] with Carpenter and Savage, and after some trouble
made it fast.

A bald enough record, but it brings memory of some lively times
handling gear.

[5] The topgallant mast was the third mast from the deck; immediately above it
were the topgallant sails with the royal sail next, and the highest sails on any
square-rigged vessel.

Lighter moments are recorded as follows:

In the last dog watch we had fine fun. We all went up on the fore-castle and formed a ring for dancing. Jack Greenwood danced the Lancaster reel. Billy Little danced a double shuffle, and Pat Keating, the fireman, together with the big messcook "English Bill," gave us a fine spectacle of a regular Irish Reel. I laughed like fun. September 15:—nothing of importance happened except that Rodgers and Butts had a fight on the forecastle. They were both pretty well pounded up. I stood a trick at the wheel last night from two to four, and the old thing kicked like a horse.

We were in the tropics by now, and the lightest of garments were the rule; barefoot always (our feet were as hard as leather from climbing tarry ratlines),[6] loose duck pants, a singlet, and a cap against the sun were our costume. The decks were furnace hot, and often pairs of us would repair to the head and douse each other with sea water. One day, a dead calm, a middy lost his cap overboard and promptly jumped after it, to be retrieved at the end of a tossed rope after a refreshing swim, the transgression winked at by officers rather envious of his ingenuity.

As we neared the home coasts we were cheered by signs of land. A kingfisher lit on the foreyard, and a hawk in the maintop where I was working; a three-masted schooner, the first we had seen since leaving American shores, gave us a glimpse of her familiar and graceful lines. Off Bermuda, we met the usual tricky Hatteras weather, and I mention being nearly taken aback in a shift of wind, when the commands came, "Settle away royal halliards; clew down and clew up! Clew down topgallant yards!" but the to'gallant halliards jammed in the sheave, and several of us jumped aloft to clear them. The squall carried away the jib sheet, and the slatting of the big sail threatened to take the fore royal mast with three hands on it, but prompt work alow and aloft saved the day. I can hear that wind howl and feel that mast shipping yet, and admire the youngsters, myself included, who stuck to it until all was snug.

On October ninth, thirty-one days out from the Canaries, the long streak of Cape Cod sand hills appeared on the port bow. "Everybody was happy," said I, "Mr. Forster came forward with a beaming smile and remarked that it was dirt, sure enough.

---

[6] Ratlines were small three-strand ropes used as a rope ladder on most ships.

Mr. Osborne gave sign of his joy by inventing a few extra jobs for the watch on deck. The 'old man' looked as if he had just got a check for a thousand, while the fellows began to smell the apple pies and oyster stews of Boston."

There is not much left to tell of the voyage. The breeze freshened, and as if to show that the Atlantic was not through with us yet, carried away the main topmast staysail, which was got in and stowed with much trouble, and under shortened sail we rounded the cape and anchored under Long Island next day. Bundled to the ears in the chilly October winds of Boston, we "scrubbed paintwork all the morning, and in the afternoon holystoned the gun deck," a fitting sequel to our arrival in Southampton three months ago.

I have to finish my tale of the *Enterprise*'s first cruise with a happening I now regret; a small mutiny, for a cause which now seems trivial but then seemed important, in which I fear I was one of the ringleaders. The powers that were, at the head of the School, decreed that we should not be granted liberty for a week or more, while being examined as to proficiency. This seemed to a lot of youngsters, hungry for home, a base imposition, and after an indignant meeting that night we nearly all refused to "turn in" in the morning (a sort of early day "sit down" strike).

The Executive promptly and properly put us under arrest. When the commander returned from shore, I with others was summoned to the cabin and given a well-deserved wigging, of which the memory still makes me uncomfortable. But, understanding as ever, he interceded for us, and that night we were granted the longed-for liberty. The "mutineers" were undeservedly made somewhat the heroes of the occasion; but as my Log says, "I think that if it hadn't been for the Captain's good nature, we would have been in the soup. One of the fellows remarked, 'Boys, it worked this once because Captain Merry is a gentleman, but the next time it won't, and it's a thing I'll never try again.' (Sound reasoning.) The next morning hammocks were piped, scrubbed, the decks washed and holystoned, and by seven bells everyone was ready for shore. A queer crowd we must have looked with bundles of clothes, curios, blankets, etc. I steered straight for home and didn't stop till I got there."

I had not lost my taste for the sea, far from it, but the prospect of school aboard a moored ship all winter was not in my reckoning, and as my father agreed that a strictly nautical education left something lacking, he got my discharge and entered me in Hopkinson's private school.

Brief as my first sea experience was, it gave me something of value that has been lasting. From a rather puny boy, it changed me to a tough, wiry youngster who could hold his own weight with an equally tough crew. It gave me a legacy of calloused hands and feet so hard that for years I could scratch a match on my bare sole; a knowledge of rope lore so that today I can't coil a rope the wrong way or tie a granny knot, and many a time have found my acquaintance with knots and splices invaluable. Best of all, it gave me an early acquaintance with boys and men of many walks in life, whose point of view I learned to understand, and a respect and liking for hard manual work that has made me scorn an idler.

Where that crew has gone after forty-four years, who knows. Probably hardly a third of them followed the sea. Some of course are dead. I heard of one, an officer of an armed liner which sunk a German submarine; Maguire drowned in harbor; another had a plantation in Brazil; my younger brother ran across Warner, one of my companions in Portsmouth, as mate of the barque *Onaway*, years later; and so on. They were a good lot; God rest 'em.

II

Vagabond

Back in school, I found it a far different affair than before. Satisfied for the time being by six months at sea, I found congenial friends and renewed interest in school activities in and around the pleasant mansion on Chestnut Street. "Hoppy" was a born headmaster and teacher. To him I owe perception that Latin and Greek were not merely chores but languages of beauty and strength. He had the faculty of humanizing the subject so that the authors and events became familiar people and things. My father, whose college education had been curtailed by the war, and who had educated himself in the classics, post war, had a great respect for them and saw to it that I got plenty of them through school and college. I have never regretted it; have realized the groundwork they have given me in the use of language, and to this day take pleasure in those passages of the poets that I remember, my favorite being Horace's ode beginning "Vides ut alta stet nive candidum Soracte," which has a homelike touch to one who has lived much in the mountains and the snow.

In '94 private schools preparatory to Harvard in the Back Bay district were very much the thing. That part of town was still a good deal of a village, and we made the Common and the streets and backyards of the "Hill" our playground. We had a field on somebody's country place out near Longwood, on which we played football, into which I plunged with alacrity. In winter, our favorite sport at recess was running "tiddledies" on the Frog Pond, which consisted in bounding from shore to shore over shaky ice, the shakier the better. I developed a kind of half run, half shuffle, which was remarkably successful in keeping me on surface, but more than once a large part of the school fell in and had to be dried out by the

21

school furnace. Another sport, not so harmless, was snowballing "rubadubs," covered grocery sleighs, as they passed down Beacon Street. They made a delightful sound, and to our minds the practice was justified, as it developed the often lacking sense of humor of the drivers. One fateful day a large and imposing citizen remonstrated, and I bustled up and asked him what business it was of his. "I'll show you," said he, and promptly seized me, showed a police badge and marched me across the Common, followed by the whole school in pairs, whistling the "Rogue's March" in time to his ponderous tramping, to his great annoyance. I was booked at the LaGrange Street station and turned loose with a warning, to the loud cheers of the assembled school. It seems that the officer was no less than the captain of the precinct. It happened that the police had been in for a good deal of criticism due to laxness of arrests lately, and a wag of a reporter made much of the incident, extolling the captain for his arrest of "that dangerous criminal, Bobby Livermore."

Springtime set up its familiar call, "Let's be off somewhere"; school was nearly over, and promptly I put aside all thoughts of boyish pursuits and sniffed the sea breeze. A Cape Horner still in mind, I searched the piers and haunted the shipping offices for an outbound square-rigger, whose every rope and spar I knew; but none was in port and at last I had perforce to sign articles on a coaster whose captain happened to need a hand. He drove a hard bargain and signed me on as an ordinary seaman at the munificent wage of ten dollars per month, but little I cared. I was at sea again. Home I went, threw a few clothes into my sea bag, clapped on a round sailor cap, bade a hasty farewell to a resigned family and departed, passing en route a few of my schoolmates, who, already far from my thoughts, got nothing but evasive answers to their questions, and no doubt ticketed me as a bit queer.

The *Timothy Field* lying at Marquand's wharf, East Boston, was an old-fashioned high-sparred two-master of 260 tons burden. On her deck stood the mate, an elderly man, bearded and kind of eye. "I'm a new hand," said I. "All right, son, cast your dunnage in a for'ard and pick your bunk, captain's at dinner." The latter soon emerged, picking his teeth, and looked me over. "What'll I call ye," said he. "Bob," said I. "All right then, Bob, get your dinner. You'll find enough on the cabin table. Tomorrow the rest of the crew'll be here and we'll get under way. Meanwhile, the mate'll give you something to do." And the mate did. All afternoon I replaced worn ratlines, familiar and not unpleasant work, perched high in the rigging in the cool summer breeze.

In the evening the rest of the crew came aboard, two Norwegians clad in sailor's shoregoing blue with visored caps, one a stocky man in his thirties, sailor written in every line, the other a lanky youth of twenty; and, shortly after, the cook, a young Negro black as the ace of spades, white teeth gleaming in a perpetual smile. The crew was complete.

I was to find life as a foremast hand far different than as an *Enterprise* cadet, where we were all youngsters of the same stripe, and where there were twenty men to haul a rope or furl a sail. This schooner existed for business alone, where the fewest hands possible were employed, and everyone must do his own job and help with the others as well when all hands were needed. The captain, named Leighton or Lakin (I never did get his name straight) was a taciturn, hard-bitten down-Easter, a driver, and a trifle scornful of my school-ship training, but when he found my lesson had been well learned, confined himself to an occasional jibe. The mate, known only as "Mister," was kindly always, and I think had a fatherly liking for me, and showed me many a trick of the coaster trade. Sea custom kept him from being too friendly with the crew, but often when we were singing or skylarking on the foredeck after work I would see him lingering near with rather a wistful look in his eye. Chris and Charley proved good souls, the better type of Scandinavian sailor then filling every American forecastle. My status was something of a puzzle to them. They couldn't figure out why a person like me would take to a forecastle voluntarily, but manners existed forward and curiosity was politely suppressed. Later on, they worked it out that I had got in some trouble at home and was lying low until it blew over, which settled the question and brought full acceptance of my presence. The cook was a character, a West African mission-trained Negro, jolly, fond of hymns or any song but with not the slightest ear for a tune. His quarters were the galley under the same roof as ours, whence we could hear the rattle of pans and cheerful sounds of embryo song, followed by a rap on the scuttle and a cry of "Come git yo supper, sailormens!" Each of us owned a tin cup and pannikin which we presented at the galley door, received the meal in bulk, retired to hatch or bunk and ate with the aid of our sheath knives which each carried at belt, for forecastles supplied no tableware.

Dawn saw us astir, bending new halliards, overhauling the running gear, setting up rigging, and the thousand and one things to be done making ready for sea; then down harbor in tow, where as I watched Boston's red brick buildings and church spires receding, I felt again the thrill of the outward bound, this time a seaman before the mast earning my pay, even if it was only a coaster. We sweated up the big sails, one halliard at a time, cast off our tow, and with a fair fresh wind passed Boston Light. Mine was the first trick at the wheel, a tryout, I suppose, and sharply watched by the skipper. The schooner was light, the breeze brisk, and I found her a handful. Much to my chagrin, I couldn't keep her within four points of the course. Somewhat to my surprise, no comment was made by authority, but when the others had had their turn with no better success I felt consoled. It seems a light schooner with the wind aft will "yaw" with the best of steering.

Calm fell with the dusk and we floated idly ten miles off Thacher's Island, whose twin beams made paths of light to our side. The day's work was over and content filled my soul as I leaned on the rail, puffed my pipe, a recently acquired accomplishment, and did not envy my brothers and sisters at home behind the twinkling lights of the North Shore.

Three days of light airs brought us to the mouth of Penobscot Bay and anchorage. Then, before a freshening breeze, up that broad and beautiful river, then a narrower channel where sheep grazed on rounded hills to the water's edge and trim white farm houses stood in sheltered hollows, to the head of a little bay where were scattered houses and wharves. Down came fore and mainsails, and under a jib we drifted to a pier and made fast. This was a hamlet near Sedgwick, our destination; its product, paving stones, our cargo.

In the morning hatches were opened, the fore boom triced up, and a long sluice rigged from hatch to shore. Then came a steady line of ox teams bringing pavings which were dumped in the hold with a roar. Our job was to stow them away from the growing pile under the hatch to every corner of the ship. It was hard work in that hot, dusty hold. The first day was a nightmare to me, not yet hardened; and, aching in every bone, my hands blistered from the rough stones, sleep was a stupor that night. A friendly teamster next day made me some leather palms; my sore muscles tuned up; and soon I was holding my own, but I can hardly look at a paving stone yet in Boston streets but I think of the seventeen thousand we loaded and towed in those uncounted days.

Our one dissipation in that port was a country dance in the schoolhouse to which the whole community went. A fiddle and an accordian made the music. The dances were all "square," waltzing and such being unknown in those fastnesses, and I was rather out of it, Papanti's dancing school not having taught me the "do se do" and "sachez all" variety, but the Andersons went at it with gusto. Frequent visits to a hidden supply of hard liquor brought that dance to an abrupt end with a rousing fist fight, more or less free-for-all, and the ox teams were slow in arriving next morning.

Our ship, once high-sided, was now low in the water. Hatches were battened down, moorings cast off and sail made, but a dead calm prevailed, so we had to kedge out of the bay. This slow and painful process consists of rowing the anchor out a cable's length, dropping it and heaving the ship up to it with the capstan, and repeat. We clawed our way out of the bay by miserable inches until the main channel was reached, when a fair breeze and tide caught us, and we catted and fished the mudhook with a sea blessing.

Once outside the Penobscot, we lay becalmed again under a heavy sky. I had turned in and was fast asleep when blows on the forecastle door and the call "All hands up, be lively now!" brought us all on deck, to find wild confusion. A sudden flaw from the north with weight of wind behind it had caught us with all sail set and threatened to tear the masts out of the old schooner or drive her bows under. Already, each surge was sending a torrent of water over the head to put aft in a boiling flood with the upward lift. Our forecastle was knee deep in water, and muffled sounds of crashing tinware and African ejaculations betokened trouble in the galley. Topsails had been dropped and were thundering away at the caps. Chris was out on the bowsprit lost to sight in the pitchy darkness, and every now and then smothered in the seas into which she put her nose, but steady seaman that he was, managing somehow to stow the wildly billowing headsails. Charley, the cook, and I manned the main boom, first to reef, then as the wind steadily increased, to furl the mainsail, then aloft to stow topsails, while the captain and mate tended halliards and wheel, and at last, all snug, squared away to run before the breeze under one job and a close-reefed foresail. I was put to pumping, as the heavy cargo had started the old timbers to leaking, and it wouldn't have taken much extra weight to have sent us to the bottom. This pump, a one-man affair in the waist, was kept continually at work by one of us until the blow was over. The job for me in the height of the gale was none too pleasant, for at every roll water would come over the bulwarks and cover me waist deep. Once, I glanced over my shoulder to see a dark wall of water tower over the rail, just in time to grab the pump handle and hang on for dear life. Down it came, foamed over me neck deep. Away came the handle in my grasp, and away went I, to bring up with a crash against the lee rail, inches from the top, only to have the breath knocked out of me by the pump box, which had fetched loose and followed me. As I lay there stunned and half-strangled, the captain ran to the poop rail and called me. Getting no answer, I heard him say, "My God, he's overboard!" but breath returned, and half-scared, half-mad, I crawled back to the pump and resumed my labors.

The night wore itself out in time but not the gale, which drove us before it all day and another night, safe enough now that the emergency was over, but half-awash, and all of us glad to turn in "all standing" wet to the skin whenever relieved from pump or wheel. At dawn the second day we found ourselves in a thick fog with the wind dying, somewhere off Cape Cod. Around us on all sides we could hear the dreary note of foghorns, and here and there the sound of a ship's bell at anchor. Not daring to risk passage of the shoals, we let go both anchors and rode it out. Every few hours, foghorns signalled the arrival of newcomers, first faint, then right among us, still unseen, to be greeted by a clamor of bells of the assembled fleet, until the roar of cable through hawse hole gave notice that the stranger had anchored too, in seas too crowded for comfort.

For six days we lay in that foggy open sea anchorage, our bell sounding day and night. Only an anchor watch was kept, so though we were busy enough by day we had leisure in the dog watch to gather by the forecastle door and smoke and yarn, Chris and Charley drawing on their sea experiences and their life in Norway for many a tale, while I, less travelled, contributed my share in song. While not boasting much voice, I had an ear for a tune and a large repertory of ditties picked up at home and at sea which seemed never to tire my shipmates. My talent had one practical use; the cook was a most appreciative listener, and begged me to teach him my songs. With an eye for a trade unusual for me, I bargained with him that for each verse taught we should get one piece of apple pie or other cabin dainty, which was accepted, and I remember that the "Swanee River" netted the forecastle quite a feast.

Charley and I used to spar with towels for boxing gloves, a sport at which I could show him a few tricks, though he could have bested me in a rough and tumble. He was a great admirer of my ordinarily well-kept teeth, having only a few snags of his own. I guess he had never seen a toothbrush before. One day I found him brushing his stumps with my brush, innocently enough. "Makes 'em feel fine, don't it, Bob," said he brightly. "I been trying to get dem in good shape for some days now, an' I t'ink dey look better already." "Keep the brush, Charley," said I, "Mine are in pretty good shape and I don't need it anymore." Thenceforth, I used a makeshift.

All things end, and one night the fog lifted, the stars came out, the bells were stilled, and all around us shone the riding lights of the fleet at anchor, the nearest not a stone's throw away. The morning sun showed us no less than sixty sails, all schooners big and little, Pollock Rip Lightship not far to windward, and six miles to leeward the low dunes of Cape Cod. Already there was a scurrying of the fleet to make sail and slip away from such a dangerous neighborhood. Joyously we broke out the sails from their stops, hoisted them, and manned the capstan to heave in the endless lengths of cable, the clank of the pauls keeping time to the barbaric chant of our African. Every rag set, we were first away, but by noon every ship had passed our deep-laden bluff-bowed hull. Once around the cape, we had to beat against the westerly wind all day, as I well remember, for my special job was climbing the fore shrouds and clearing the tack of the topsail every time we came about.

The rest of the voyage down Vineyard and Long Island Sound was one long monotony of head winds, calms, and tides which set us back as much as we had gone forward. A tardy breeze finally set us in at the "Narrows," where after some sharp dickering a tug took us in tow through Hell Gate to anchorage off Newark, my second and far different arrival in New York Harbor.

Learning that the voyage for which we had signed articles was to end here, and visualizing those seventeen thousand fifteen-pound pavings to be handled again, I asked the "Old Man" for discharge. Said he, "Get a pot of tar and tar down the main riggin' first, and then I'll see 'bout payin' ye off." So, tar it I did, and so thoroughly that my hands for weeks afterward were the wonder of my friends at home, a parting memento of the *Timothy Field*.

My small pay in hand, I boarded a train for Boston, sailor cap, sea bag and all, and must have been a picture of the "sailor home from the sea," judging from the amused looks of my fellow passengers. I know that my father welcomed my return, for the storm had been reported as quite severe, and nothing had been heard of the schooner for some weeks; but parents' worries rest lightly on their children's shoulders.

So ended my experience with a professional sailor's life. I still liked the sea and always will, but I realized that the romance of sailing days (if indeed it ever existed) was over.

III

Sports

Although my passion for a sailor's life subsided, my liking for outdoor life in general continued, and home thereafter saw little of me when vacations allowed me to get away on some expedition or other. Having little use for the forms and exactions of polite society, and being rather "girl shy" to boot, I had until long later a somewhat one-sided acquaintanceship, mostly with male friends of similar tastes. That remedied itself in time.

Except for football, which I took to like a duck to water, I was never very good at organized games. I liked the open country and the sports that went with it. I used to think that if I had a dog, a gun, a canoe, and a horse, I would have about everything needful in life. All of those possessions became mine in time, but, needless to say, I have found a few other things essential.

I learned horsemanship first from father, who kept his wartime liking for a horse for a long time, but I soon found bridle paths tame, and struck out across country whenever I had a chance. We boys usually had a pony between us, but sometimes the family carriage pair were requisitioned, and "Rambler" or "Vicksburg," much to their indignation, were urged from their accustomed trot to a hard gallop, and an occasional pop over a stone wall. I imagine father often wondered how the pair came to bark their knees in the stall so much.

Horses have been a large part of my life, both in the East and the West, and would be yet, had I the time and money, but automobiles and planes have made the earlier long horseback journeys of a mining engineer needless, and at home, where formerly the rich had

29

autos and the moderately endowed, horses, now it is the other way about, so I drive a Ford.

My sister and her husband, Bulkeley Wells, had a delightful farm in Needham, and, among other things, a string of polo ponies, who served me as hacks or hunters as the occasion arose. They gave me a 13½-hand pony called "Gold Dust," who could climb anything she couldn't jump. Another one, a tough buckskin, I rode by invitation in the Norfolk drag hunt, who took every jump and finished among the leaders, though the others took fresh mounts at the half-way check.

This was then a riding country par excellence, and I and my friends were cross-country minded. We had a Sunday morning ride from one hospitable country house to another, called the Scotch Whiskey Route, in which the rule was that no one could ride more than fifty yards along a road. The results of three or four stop-ins were sometimes spectacular, but alcohol mixes better with horseflesh than it does with gasoline, and I remember none but laughable casualties.

I visited the Thomas Hitchcocks in Westbury, Long Island, with my best of friends, DeLancy Jay, where after trying me out on a safe hunter and finding my seat and hands good, Mr. Hitchcock entrusted me with his best. That kind of riding was a revelation to me. I found that soaring over a five-foot rail on the back of a thoroughbred was far easier than buck-jumping stone walls on a pony. I can still see in my mind's eye the gallant Mrs. Hitchcock putting her horse at a lofty rail fence on an uphill grade, which to me, following close behind, seemed madness. She cleared it nicely and so did I. I had to. Grand people these.

Hitchcock was a sort of patron saint of the American foxhound, as opposed to the imported brand. To encourage the sport, he offered to ship us a pack of four couples. They arrived in a freight car in Dedham, and Arthur Rice a sport-minded friend, and I on horse met them. Thinking to marshal them at heel, we ordered them turned loose, when all eight promptly scattered over the streets of Dedham, two in full cry after a cat, one down the tracks, the others here, there and everywhere. We finally collected two of the more docile and two more were found later. The others vanished. This was the nucleus of the Westwood hounds, which hunted drag, fox, rabbit, or cat, whichever took the fancy of the pack.

I liked shooting as much as riding. The wily partridge was my favorite game and still is. John Saltonstall, another best of friends, introduced me to duck shooting at Cape Cod, where we spent many a holiday in the little shack on the beach opposite South Orleans, with our native Cape Cod friends and guides, Eli and Oscar, whose sayings would fill a book. Because of market shooting, and the fact that the Cape is off the line of the big migrations, ducks were none too plentiful, probably no more so than today, but we had more chance at them by reason of no bag or hour limit. Moonlight nights, now barred, when we merely set out bunches of eel grass as decoys, gave us the best opportunities. There was something deliciously mysterious in the whistle of unseen wings over the silent marsh, and the sudden loom of black shapes over the decoys. I can still hear Oscar whisper, "Here they come; Bob, here they come; Naow! put it een to 'em!" Or, at the "pint of the medder jist dark" to see against the last glow of daylight the specks of the incomers from the "rafts" outside, sometimes to our decoys, more often not.

Partridge shooting was a simple affair of boarding a Saturday morning train whose smoking car would be filled with canvas-jacketed devotees, their dogs sprawled over the seats and floor, and going some twenty miles out. In the day's walk we would be sure to see twenty or thirty grouse and a covey or two of quail. The experts with their well-trained dogs would bag a half dozen or more, but I with no dog or one half-trained, counted it a good day with two or three. My first setter was good at finding and pointing, but if the bird went down it was a race between him and me. If I lost, I usually found the bird half eaten.

That most admirable of game birds still furnish good sport in spite of epidemics, vermin, and many times the number of hunters, but the covers left around Boston harbor one bird where there were many. I believe, in spite of argument to the contrary, if the same number as before are seen farther afield, it is because the automobile has made it possible to cover ten times the country. The wonder is that there are as many as there are. Good old Bonasa Umbellus; I like everything about him to shoot, to eat, but most of all to see him alive and flourishing.

I first met big game in the White Mountains of New Hampshire, where as schoolboys two of us used to go during our winter vacations to stay in a little farmhouse on the Dundee Road near Jackson. Life here was lived in primitive fashion, much as these people's forebears must have lived a hundred years ago. A central stove was the heating plant, candles and a lamp or two the light, with plumbing of the outdoors variety. With the temperature well below zero, it took courage to step out of the big feather bed on the bare floor in the morning.

The sons of the house were backwoodsmen, knowing every trick of the woods and its inhabitants. We stalked the big white hares to their forms under the spruces, shot partridges budding in the birches against the light of sunset, and set traps for foxes or ran them with the lanky hound.

One day in midwinter two of us donned snowshoes and broke trail through a four-foot-deep snow up the side of Mt. Kearsarge. In a thicket of hardwood, high up on the mountain, we found what we were looking for, deep-trodden trails of deer in their winter yard. Patiently "Fred," native son of the mountains, worked out the tracks in the mile labyrinth of paths, gradually pushing the deer ahead of us, until they were in a cul-de-sac and had no refuge but to plunge through the deep snow. I saw the first one, a doe, showing brown against the snow, unable to go farther, and killed her with one shot. As it was late, we gave up the chase for the day and dragged her down the mountain to a cache near the road. The next day, eager for more, we climbed our trail again and took up the chase. We found where the deer had left the yard in desperate plunges and each took up a track. Mine, soon overtaken, was a big buck, first sighted where he stood under a fallen log. My shot only wounded him, and a second was out of the question, for my rifle was plugged with snow. I drew my hunting knife, and snowshoes on, jumped from the log to his back, to find myself in a real tussle with a very lively, powerful animal. It was a question which would win, but I finally finished the bloody and distasteful job. Fred had downed his deer also and we spent the remaining hours of light towing our quarry to the hiding place from which we retrieved them all next day.

It was cruel and unsporting, looking back on it, to say nothing of being in total disregard of game laws, but I was young and but followed the custom of the country. Whatever compunction I had was stilled by the thorough use to which the windfall was put, and the real need of these people, to whom money was almost non-existent, and fresh meat a rare luxury. Not a thing was wasted, from the hide to the little ankle bones, which made lasting toothpicks for the family.

I think I and a few others in Jamaica Plain were the first to ski around Boston. We had long, ungainly affairs made by a local Scandinavian carpenter, with no bindings but a toe strap and a block behind the ball of the foot. On these, we got fairly expert at straight running and jumping off walls, but of course had no control in the modern sense. I have kept it up all my life, following the various changes in rig and technique, and find it one of the finest sports, even at my present age. I take some credit at having put my son on skis at the age of six, which may have started him on his way to land among the top-notchers in downhill racing.

At football, the one game I played well, I developed a faculty for getting around the ends, which resulted in good gains for my school team and brought me to the attention of Harvard scouts looking for schoolboy material. As they played freshmen in those days, I was called out for the spring squad even before I had entered college. I spent most of that summer learning to kick under the tutelage of a past Harvard star, Jack Fairchild, at Newport. The Fairchilds lived a delightful, affectionate family life in their house on the old waterfront of Newport, and made me a part of their circle in warm and friendly fashion. Why so fine and happy a family should in after years suffer one tragedy after another has always been a mystery to me.

Football was an especially tough game in those years. The "blood and iron" policy was traditional, the theory being that the bigger gruelling you got the hardier you were. It was the day of light backfields and mighty linesmen, and my only salvation as a halfback weighing 140 pounds was to get tackled as little as possible. I guess that fear had as much to do with my agility as skill. As it was, I got so smashed up in my freshman year that I wasn't much good afterwards. I made my letter "H" that year, mainly because all the first-string men were used up, but after four years of getting more and more bruised and stale, I ended my football career as an obscure substitute.

The summer of '95 I had my first trip in the northern wilderness. Four of us, Farley, Cochrane, Sherwin, and I, camped for a month in the Squatook Lake country of eastern Quebec. There, we learned how to handle an axe, to cook camp fashion, and how to pole a canoe in swift water. We fought incredible swarms of mosquitoes, and caught trout until we were tired of eating them, but saw little game. On return to Edmundston, we turned loose our only and nearly useless guide. Cochrane went home, and three of us set forth in two canoes to pole our way up the St. John, and by way of the Allagash River through the heart of the Maine woods to Moosehead Lake.

It took three weeks of strenuous work against rapids to make the trip. Eight miles a day was the best we could manage, but it was a fine experience as training for muscle and woodsmanship. Once in the Allagash, we found ourselves in a game paradise. Hardly a day passed without a sight of deer wading the stream, or a moose breast deep in lily pads, and we lost no time in supplying ourselves with venison.

I don't know what the Allagash is like now. I hope it isn't spoiled by too much travel and policing like most American wildernesses. It was always a thoroughfare, but a woodsy one. The loggers had taken the best of the pine and gone, leaving the woods in peace for a space; passing canoes were few, and of supervision there was none. I have often wondered why there was so much game, where there was little or none in equally wild Quebec. Certainly it wasn't because of strictly enforced game laws, as there were few wardens (we saw none) and the traveller helped himself to meat as he needed it. Thoreau, in *The Maine Woods*, mentions the sight of game as a rare event. Perhaps ours was an extraordinary year.

On the strength of having made the trip, the following summer I volunteered to guide two friends, Mott Shaw and Joe Bigelow, on a return trip downstream. This voyage was a story of mild disaster from the start. First, we folded one canoe around a rock in Pine Stream Falls of the West Branch of the Penobscot; bought another at a settlement on Chesuncook, a waterlogged veteran of many coats of paint; then, trying to run Chase's Carry, five miles of wicked water, upset the canoe carrying most of our provisions and recovered little. We knew there was a loggers' supply camp on Long Lake, at which we hoped to replenish our much depleted stock of food, and decided to make for it rather than turn back. We were a pretty hungry trio by this time. As we paddled down the long stretch I noticed Shaw looking back every now and then. Afterwards, I asked him why, and he said: "I felt as if there was something following us." "There was," said I, "It was the wolf of hunger."

At the camp there was nobody and no provisions in the log house, but there was a flourishing field of potatoes into which we dug, and thenceforth they formed our *pièce de résistance*.

This happened to be a rainy summer, and except for a caribou cow and calf, which incidentally must have been among the last seen in Maine, and a rare deer or moose, we saw little game. They probably found plenty of water and feed in the smaller lakes. At any rate, we did not bring any to bag, and by the time we reached the St. John River and farms, we were ready to eat anything, barring potatoes.

My recollections of that trip are mainly of soggy camping places, unburnable wood, continuously wet clothes and bedding, and never enough to eat. When Bigelow, in haste to add to the larder by shooting a duck, shot a charge of bird shot through our one remaining good canoe, causing a forced camp for repairs, on a diet of potatoes, our cup was full to overflowing with a reverse English.

I was not proud of my record as guide or woodsman on that trip.

I V

West, Young Man

In '98, when I was a sophomore, came the Spanish War, and of course I wanted to go. An opportunity came when Roosevelt, then in Florida with his Rough Riders, sent for a contingent of six, and I was one of those picked. I went home to say good-bye, but father persuaded me that I would do better to wait until fall, so as not to sacrifice my college standing. Then, if it were to amount to anything, he would give me his blessing and aid. As it turned out, if I had gone, I would have spent the war in Tampa with the others of the contingent who did go. My brother Harry,[1] who was rather at loose ends at the time, joined the First Engineers, and campaigned in Porto Rico, seeing no fighting, but nearly dying of typhoid, and coming home the thinnest soldier I have ever seen.

Father, knowing my outdoor tastes, wanted me to try my hand at the mining business, and got me a chance with a small exploration company having mines and prospects in Colorado and Wyoming, of which Bulkeley Wells was in charge. I was as interested in this, my first trip west of New England, as I had been in my first trip abroad. Trains were slower then and I had ample time to observe the change of scene. My diary records my impressions of things new to me, from the start. Birds were always of interest to me. I made note of many flocks of doves, and of redheaded woodpeckers seen from the car windows in the midwestern states, and later, of prairie chickens, all now rarely seen from the train.

The Great Lakes, inland seas, awoke my sailor instincts, and in

[1] Two years older than his brother Thomas, Harris Livermore was in the import-export business in New York for several years. He was killed in a plane crash in Mexico in 1929.

Chicago I wandered to the docks to see the shipping. There were steamers and plenty of schooners, one of which with a square fore-sail was putting to sea. A mate standing alongside his schooner asked me if I wanted to ship, offering good pay and grub, but I, with deep-water scorn, said, "Not on fresh water, thank you."

I awoke in western Kansas to find the train speeding over a flat green prairie, no fences except around a very occasional ranch, a dim wagon track for a road. Once, I saw a pair of cowboys sitting on their horses, and once two antelope paced the train at a few rods' distance. This was still a cattle range, before the day of the dry farmer, who now has fenced the land solidly, broken the sodded grama grass, ekes out his scanty living from crops in the intervals of drought and dust storms, and calls to high heaven and Washington for aid.

In Colorado Springs I joined Wells, spent some pleasant days in that oasis of eastern culture, and explored the surrounding plains ahorse. We visited Cripple Creek, then six years old, a typical mining camp in its early bloom, but our stay was so brief that I got impressions only of puffing hoists, six-horse ore wagons careering down narrow mountain roads, felt-hatted miners, bearded prospectors with their loaded burros, and the dull boom of underground blasts. I was to see a lot of "Cripple" in later days.

Thence we went to Wyoming, took a wagon at Laramie and drove fifty miles into the Medicine Bow Mountains to visit a placer mine[2] which Wells promptly shut down (no "pay"), thence by train to Rock Springs, and two days by wagon to our destination, South Pass City, one hundred miles from rail.

All of this was full of interest to me in its total change of scene. The rolling fenceless range, the new bird and animal life, the incidents of the trip, such as the little lake where we camped for the night, where wild horses came down to drink and fled in panic at the sight of us, a wolf which stood, a lone sentinel on a rocky hill, and finally the small log-built street of South Pass, where once the Forty-niners

---

[2] A placer claim was technically a mining claim located on top of the ground in which the ore was obtained by sluicing or washing. Although in practice there were numerous interpretations as to size, in legal terminology the placer claim size was defined as 1,320 square feet (Albert H. Fay, *A Glossary of the Mining and Mineral Industry* [Washington: Government Printing Office, 1920], p. 517).

passed, now a remote settlement supported by a single gold mine, the Carissa.[3]

Our outfit had a prospect near this mine, and were driving a tunnel aimed to cut the extension of their vein, which incidentally it never did. We had as crew a couple of miners, an assayer who set up his shop in an old cabin, Blackmer, the engineer, and I, who was installed as timekeeper, cashier and general assistant. As we had no ore to assay, and as payday came but once a month, my duties were not arduous. I would have found time hanging heavy had I not been enthralled by the western scene, and eager to learn its ways. I explored the surrounding country, shot sage chickens, chatted with sheepherders and woodchoppers, rode with Blackmer to the neighboring semi-defunct mining camps of Lewiston and Miners Delight, where we examined abandoned mines at the risk of our lives,

---

[3] Eight miles southeast of present-day Lander, Wyoming, South Pass City grew in response to the needs of emigrants who came through storied South Pass, a smooth, low geological saddle, which was first discovered by Robert Stuart in 1812 and soon became the main gateway to the West on the Oregon Trail. The only substantial growth of South Pass City came with the mining boom in the late 1860's and early 1870's (Carter County records, Western History Research Center, University of Wyoming Library).

The Carissa lode was discovered on June 8, 1867, by H. S. Reedall. By the next summer there were approximately two thousand miners scurrying over the South Pass region in hopes of hitting another Carissa. The first stamp mill (a building wherein ore is crushed by pestles powered by water or steam) was constructed in July, 1868. The second stamp mill of ten stamps was built by July of the next year. A reputed sixty thousand miners and various other persons inhabited the South Pass area in 1869. All the principal mines were in operation by 1871. Four years later the mines were dead for all intents and purposes. During a brief episode in 1886, the Miners Delight district was worked. A French company, purchasing a group of placer claims in 1884, attempted to increase their production by hydraulic mining; but after several lean years they sold out to the Dexter Mining and Milling Company, whose success did not exceed that of the French capitalists. When Livermore visited South Pass, the Dexter company was finishing operations and seeking its fortunes elsewhere (*Sweetwater Miner*, March 24, 1869; C. G. Coutant, *The History of Wyoming from the Earliest Discoveries* [Laramie: Capalin, Spafford & Mathison, 1899]; *Wind River Mountaineer*, May 18, 1889; Arthur C. Spencer, *Atlantic Gold District and the North Laramie Mountains* [Washington: Government Printing Office, 1916]; W. C. Knight, *The Sweetwater Mining District* [Laramie: 1901]; Henry C. Beeler, *A Brief Review of the South Pass Gold District* [Cheyenne: S. A. Bristol Co., 1908]; Albert B. Bartlett and J. J. Runner, Atlantic City, *South Pass Gold Mining District* [Cheyenne: S. A. Bristol, Co., 1926]).

and soon was friends with most of the small population thereabouts.

South Pass was a sleepy place ordinarily, but occasionally came to life with a bang. Its one street boasted a store, a hotel of sorts, a couple of saloons, and a dance hall, all built of logs with towering false fronts. Local color was supplied by the infrequent visitors, now an Indian buck from the neighboring Shoshone reservation,[4] his long black hair flowing under his Stetson, his moccasined feet belaboring his pony's sides; rarely a buckskin-clad "mountain man," who, disdaining shelter, rolled down his bed and picketed his horses alongside the road; and once a dozen rugged prosectors down from the mountains with a numerous pack train. Once I met a sheriff and his posse, armed to the teeth, on the trail of a horsethief. They questioned me closely and galloped on their way. Afterwards, I heard they found their man asleep and ordered him to throw his hands up. At the same time, one of the posse—whether nervous or just taking no chances, I don't know—shot and killed him. Our citizenry, not being horse-minded, were quite indignant about it.

Prairie schooners were a frequent sight, sometimes singly, sometimes in long trains, filled with backwoodsy families and their belongings. These people, mostly from Missouri and Arkansas, were tailenders of the "covered wagon" days, filled with the same restless spirit of the pioneers, engaged in settling up the still vacant prairies of the Northwest. I talked with many of them and found them kindly, independent, and unbelievably simple. One such, eastward bound, discoursed on his lack of success in western homesteading, saying he was going back to Missouri, and all he wanted was to "see a shock of corn and look a fat hawg in the face."

As an example of how open the West still was, Blackmer and I,

4 Northwest of Lander, the Wind River Reservation was first set aside by the Fort Bridger Treaty Council of 1868. By a series of treaties in 1872, 1896, and 1904, the original patrimony was reduced from 3,100,000 to 584,940 acres. Initially the reservation was intended for the Shoshone tribe, but it is now shared by the Arapahoe tribe (Charles J. Kappler, ed., *Indian Affairs: Laws and Treaties*, 3 vols. [Washington: Government Printing Office, 1903, 1913]; James I. Patten, "Report," 45th Cong. 2d Sess. *House Executive Document 1*, pp. 1–17; H. D. Del Monte, *Life of Chief Washakie and Shoshone Indians* [Lander: privately printed, 1945]; and Virginia C. Trenholm and Maurine Carley, *The Shoshonis* [Norman: University of Oklahoma Press, 1964] and the manuscript file, Western History Research Center, University of Wyoming).

returning from a mining trip, came on a herd of eight hundred unbroken horses in charge of several riders, who were trailing them from Idaho to market in Oklahoma, camping and grazing as they went. We watched them as they drove their herd in bunches of twenty past men stationed to take tally before rounding them up for the night. These men were most dexterous with a rope. When the tally was finished, they cut out their saddle band and while others held them without aid of corral, the ropes shot out from every side and each time circled the wanted mount. I was told I could have any horse in the herd for ten dollars, but as they were unbroken, was not tempted.

The arrival of the prospectors' train above mentioned was the signal for a jollification. Among the men was an old fellow who played an instrument called a dulcimer, a piano-like arrangement played with buckskin-covered mallets. The old chap was an artist, and to cap matters one of the others was an expert "caller." The dance hall was swept out and all the miners, their wives, daughters and friends, ranging in age from ten to fifty, gathered. As in Sedgwick, Maine, all the dances were quadrilles, but with a difference that with a caller who knew his art no one could fail to do his part in the figure, and with confidence imparted, the infectious lilt of the music made us all experts at filling his every demand. For long hours the old hall resounded to the "Arkansaw Traveller," "Balance to your pardners; Gent around the lady, and the lady don't go; Lady 'round the gent, and the gent so lo!"

At two o'clock the dance ended, but not the festivities. The men adjourned to the saloon for a night cap, which, in the way nightcaps have, consumed the rest of the night. Again, the dulcimer was brought into play, this time for jigs, at which another talented prospector shone, and in which we all joined, ending up with a good imitation of an Arapahoe war dance. For refreshments, venison was brought from their camp, cut on the bottom of the versatile dulcimer, and distributed. By morning's light a dog fight, following lively boasting and betting, was staged between the pride of South Pass, a bulldog named Hot Tamale, and the strangers' best. The local talent won hands down and chased the stranger out of camp.

It was on this occasion or one like it that my enthusiasm led to a bet with a local sportsman that both of us should dive from the bridge spanning the shallow creek which flowed back of town, and

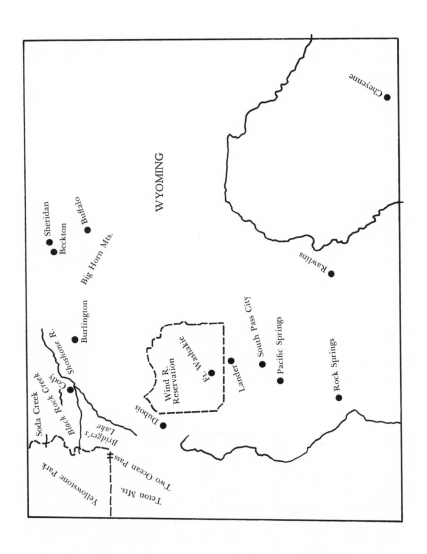

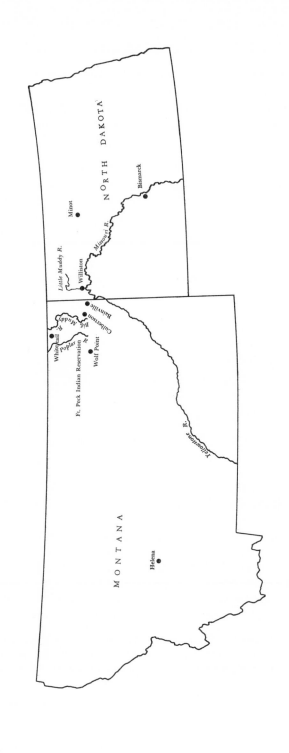

the worst dive would stand the drinks. The bridge was about six feet above the water, and the water was about six inches deep. We lined up, and at the signal of a shot from an obliging citizen's pistol, I dove, landing on my hands in the pebbles at the cost of barked knuckles, but luckily no other injury, while my opponent reneged at the last moment. I got quite a reputation by that.

South Pass was getting more crowded. Dust clouds heralded the coming of huge bands of sheep on their way down from the hills to the fall range. They surrounded the town with a sea of woolly backs, filled the air with raucous baas and acrid smell, and left nothing but trampled sage behind. Their herders came into town to establish contact with men and saloon cheer again. Miners looking for work, and drifting cowboys dispossessed of their range, rode in and added to the population. News of its growth must have reached the outside world, for one day a prairie schooner stopped on the far side of town disgorged a lean, sharp-featured man in a green sweater and a hard hat, who looked the landscape over for a camping place. Behind him from the canvas peered two rouged and ringletted females. By sundown a tent was up, and by dark a row of sheepish figures was lined up outside. South Pass had its "redlight district."

In August, it was evident that the job was petering out, and orders came to pack up. Although I had made rosy plans, with little thought for my future career, of going north to the Wind River Mountains with one Del Pratt, an attractive and roving-minded friend I had made, there to hunt and prospect, Wells' wiser counsel that I would best round out the season by seeing more mines, aided by a hint that there might be a hunting trip with him, won over Pratt's seductive tales of the mountain man's life of adventure, and I bade him farewell and engaged passage with a freighter for the one hundred mile trip across the Red Desert[5] to rail.

The freight outfit was a four-horse open wagon whose driver was an oldish picturesque individual, with one eye, called "Injun Dad" Richardson. He owned to a quarter Cherokee blood, and probably had more, a tribe which has assimilated with the white better than any, and has darkened the hide of many a plainsman otherwise

[5] The Red Desert, in southwestern Wyoming between Rawlins and Rock Springs, is an area of approximately 11,000 square miles. Whipped by searing winds, its salt-impregnated soil supports only such vegetation as greasewood and sagebrush.

indistinguishable from his fellows. We were joined by two other passengers, Bluett, a Cornish miner, and Sam Bethel, a cowpuncher on the move, with his bedroll, saddle and two good horses which he hitched behind, remarking that he was tired of forking a horse and was going to ride in luxury for once. After a few rounds of parting drinks with all the male population of South Pass, Dad cracked his whip and amid a salvo of revolver shots we galloped out of town. Pacific Springs,[6] twelve miles away, one store, one saloon and a house or two, suggested a soup for thirst quenching, a stop which lengthened to an all-night session of poker, from which after one look at the hawk-faced inhabitants, I wisely refrained, and, bored and weary, bedded down in Bethel's unwanted bed.

In the morning I routed out Dad, and helped him hitch up and get under way, but not before parting drinks were taken. On taking inventory, I found Bluett dead to the world, reposing amid the freight, Bethel much the worse for wear and minus his two horses and saddle, Dad rapidly getting incompetent from repeated drafts, four lame and weary horses and myself, sober, disgusted and anxious to get to the railroad. After several trials at driving which resulted in wandering off the road, breaking the harness and losing his whip, Dad turned the lines over to me, saying, "T'aint no use kid, you gotter take charge of this expedition." It seems funny now, but it wasn't at the time. I was still pretty green. I had never driven four horses, especially half-broken ones. The country was vast and unfriendly looking, and the road badly marked. Bluett knew nothing but mining, Bethel disdained freighting, and Dad, the only one who knew the way, remained semi-comatose as long as the dozen bottles we had taken aboard at Pacific lasted. I made up my mind I was in for a real job, and I flatter myself I was the chief cause of seeing the trip through.

There was little food, for none of us had thought to bring any, so I kept a weather eye out for game, and managed to bag a half dozen sage chickens and a rabbit, and in addition dressed and cooked them. I did most of the harnessing, saw to it that the horses were hobbled

[6] About two miles from South Pass, almost every footsore, bone-tired Oregon Trail diarist refers with relief to this cool spring and the luxuriant grass. The spring was just over the Continental Divide, with the water eventually flowing down the western slope of the Rockies into the Pacific Ocean. For a very short time during the 1880's, a stage station was located here.

and turned where the feed was best, and generally nursed the outfit along. Washington's Ranch,[7] where we camped for the night, was a bright green spot in the grey desert of sage. Here we cooled our fevered brows at the ranch-house spring and our horses feasted on green grass under the irrigating ditch. All night the coyotes kept up their high-pitched chorus, while we, after a meal to repletion of grouse à la bacon grease, rolled in our nondescript bedding and slept under the stars.

The liquor had given out and a sober outfit inspanned and started out next morning, but before many miles, we met, first, a freighter who stopped for gossip and produced the inevitable bottle, then a party of elk hunters ahorse and in wagon, who supplied us, to my disgust, with more drinks and a bottle or two for the road. Soon, the effects were felt in the driving. We were now passing down a road out from the edge of cliffy hills. Dad lashed his team to a gallop, and as we approached a bend vanishing apparently in thin air, flourished his whip and said "See that thar cliff? That's where Jim Hodgson lost his lines and went over the edge. Broke his neck plumb off, it did!" "Take them lines, kid," growled Bethel, "or that damn Cherokee'll bust all our necks too." So once again I resumed the leadership and drove the ticklish curves and the final miles to Rock Springs, where late at night we disbanded, thankful to be quit of each other.

I don't know why I tell this rather uneventful tale at such length, except that it made an impression on me as a newcomer to the West, thrown more or less on my own resources, with old-timers through circumstances dependent on the "kid" who rose to the occasion. I think the episode gave me a feeling of being part of the country that nothing else could have done as well.

I rounded out that summer by a trip or two in company with Wells and Benjamin Lawrence, his mining consultant.[8] We seemed

[7] Washington's ranch, twenty-two miles southwest of South Pass, was a favorite way station for travelers until the route to Rock Springs was altered with the advent of the automobile (Interview, Mable Moudy, July 23, 1965, Laramie, Wyoming).

[8] An entrepreneur as well as a consulting engineer, Benjamin Lawrence was a long-time associate of the Livermore family. He served in an executive capacity with several Livermore- and Whitney-dominated concerns, including the Contention Mining Company, the Colorado Superior Mining Company, the Metals Exploration Company, the Humboldt Mines Company, and the Smuggler-Union Mining Company.

to visit mines only to shut them down. Then in September we made up a party of Wells, myself and, somewhat to my scorn, of my sister Grace and her friend Maidie Sherwin, for a hunting trip in the Flat Tops of western Colorado. Let me say now that the girls made good and raised the level of their whole sex in my estimation.

It was rather a party de luxe, but not too much so. We had two guides, ample pack and saddle horses and a couple of hounds for bear or "lion." Starting out from Glenwood Springs, we camped first at Marvine Lake,[9] where trout were to be caught simply by casting any type of fly from shore, and sometimes striking three to a leader. Here I saw a big cougar who made off before I could get over my surprise and took refuge in his cave behind a ledge ten feet down from the top of a sheer two-hundred-foot cliff. Wells tried to reach him by having himself lowered by lariat but found it too hazardous even for him, much to my relief.

We then moved to Lost Park,[10] a paradise for black-tailed deer and for their hunters. We had only to ride through open quaking aspen forest, jumping does and fawns every few rods, until the buck with the right-sized antlers came into view, when with little effort at concealment, a short stalk brought him in range and down. I was not to see that plenitude of game again for some years.

So, weaned away from my vague plans of becoming a "mountain man" in Wyoming, I went meekly back to college, and the Spanish War being over, settled down to work and football.

[9] Usually known as Trappers Lake, but sometimes referred to as Marvine Lake, from the name of the nearby town.
[10] Approximately eighteen miles from Kenosha Pass, Lost Park was a wilderness surrounded on three sides by swamps.

V

Cowboy Interlude

My taste of the rougher fringe of the West had fired my imagination to the point of trying more unsupervised contact with it. From Bethel and the other roving cowboys I acquired a scorn of sheep, and I listened eagerly to tales of the real cow country farther north. One Mr. Brewster, a retired cattleman, lived near us in Jamaica Plain, and to him I went for counsel. He told me that about the last stand of the old range life was in northeastern Montana, from the Missouri River up to and beyond the Canadian border. He had an interest in the Bar Diamond brand, ranging in that section, and gave me, in case I came across that particular outfit, a letter to the boss, Frank Arnett.

I simply had to try cowpunching before I settled down to the prosy life, as I then thought it, of a mining engineer, and as soon as examinations were over I boarded a train for the West again. In St. Paul I bought myself a saddle and a pair of boots, both, as it proved later, not of the prevailing plains style, and a ticket for Williston, North Dakota, where the map showed the most vacant spaces. As the Great Northern train went westward, first farms, then ranches, then even fences, grew scarcer and finally disappeared, leaving rolling plains covered with prairie grasses in their place. When toward dusk, I saw a horseman, his bedroll broken over the pack of his lead horse, just riding into a grassy lake to make camp, I tossed my straw hat out of the window, clapped on my old felt, and thought, "The real range at last, and I am going to be part of it."

Williston was a town in process of change from pioneer days, part log-built, but with one or two brick buildings of pretention. Its main street stretched out, western style, from the station, to which at train

45

time the whole populace repaired. No doubt I was given amused, appraising glances, but, unconscious of them, I shouldered my pack and walked to the hotel, a gaunt frame structure up the street. I was assigned a room with a large double bed, and found someone's else dunnage there. To my remonstrance, the clerk replied with surprise: "You didn't expect a whole room, did you?"

My partner and I established the *entente cordiale*, but when a third individual, drunk, arrived after we had rolled ourselves up in bed, and tried to get into the same bed with his high boots on (if I remember correctly, he had spurs on too), with one accord we vigorously shoved him to the floor, where he rolled his coat for a pillow and philosophically snored himself to sleep.

Inquiries in the morning brought out the fact that in spite of appearances the range country was farther west, over the Montana line; that this country was filling up with ranches and the herds were all small. I was told that a big herd of horses from a disbanding cow outfit was at the corrals in town and I could get a good cow horse there. I got in touch with the wrangler, who told me I could have my pick for thirty dollars. He drove them around in a circle and I spotted one, a fine-looking bay, which was promptly roped and became mine. This was "D.K.," named for one of his many brands, and a good horse he proved except for some peculiarities. When first saddled, if not led forward a few steps he was apt to rear and throw himself over backward. When I tied him in a stall and fed him oats he gave one snort and set back against the halter, which broke, and over he went, with me dodging hoofs as best I might. He probably had never seen stalls or oats before. He would not buck unless ridden bareback, but then, as I found by sad experience, he unloaded his rider without delay. Aside from these little peccadillos he was a good horse, easy-gaited, enduring and reasonably gentle. I had him for all my stay in Montana.

I tied my pair of blankets behind my new creaking saddle and headed west, Culbertson, Montana,[1] fifty miles away, my destina-

---

[1] Named for the Hudson's Bay Company factor at Fort Union, Culbertson was of rather nebulous origins. "The theory that there was a town gained currency between 1888 and 1892." In the latter year one, "Lucy A. Isbel stepped off the train and spent some time looking for it. Two log buildings were not regarded as a town where she came from" (Federal Writer's Project, *Montana* [New York: Hastings House, 1949], p. 224).

tion. A long and lonely ride it was, and late in the afternoon, saddle-sore, I was beginning to wonder whether it was on the map at all, when I met a horseman who assured me that it was, and only ten miles away. Elmer Astrup, so he was named, was out hunting strayed saddle stock and for company turned around and rode in with me. He proved a friend in need, and as we jogged along told me of the country and its life, past and present. It seemed that the day of the really big cattle outfits was over. Farming had not come to any extent west of the Dakota line and the range was still open, but sheep were advancing from the western foothills and usurping it, while ranchers were fencing up the water and running only small herds. Only a few of the old cattle companies, of which the Bar Diamond was one of the last, were "running a wagon." This outfit, he said, was grazing some two thousand head of beef steers on the Fort Peck Indian reservation lying along Poplar River, a stream flowing from the north into the Missouri thirty miles west of Culbertson, and had another herd of three thousand "dogies" along the Big Muddy, farther east. He was doubtful whether they wanted hands, as the roundup was over and extras were being laid off, but he said: "You can always try, and meanwhile make yourself at home at my place." Culbertson was an old-time cow town still undiluted by settlers. It developed that Astrup was proprietor of the only saloon in town. As we rode down its one log-built street, he pointed out with pride his "place," from which, as he spoke, an individual bolted out with wild yells, followed closely by another who cut loose with his six-shooter, apparently bent on murder. I looked at Elmer with startled glance. "It's only the boys smoking things up a bit," said he, "Sheriff's away."

We turned our horses in the corral, and soon I was refreshing myself with "steam beer" and getting acquainted. The evening developed into a large one. Elmer was a virtuoso at lively fiddle music, and spurs jingled and high heels thumped the floor. I "set 'em up" in my turn, and soon was included in the company without question. Among the stunts I remember was of one who brought his horse into the barroom, saddled him up, then rode to the bar and ordered a drink for him. This was a bit too thick for Astrup, who shooed him out, saying there were limits in a man's own house. However, it was a good-natured crowd, and though the revelry went on all night and well into the next day, no one broke the peace except in fun. As for me, I turned into my blankets in a side room and slept the sleep of the well exercised.

Among the punchers and others gathered, there was one Clint Otis, a fine-looking man, who with his tanned face, his flowing mustache, bleached shades lighter, his trim rider's figure clad in white shirt, black loosely tied neckerchief, corduroys, and high-heeled boots, could have posed for a picture by Remington. I was told he was the "rep" for the Bar Diamond. A rep is one who travels to other roundups and picks up strays of his outfit's brand. I asked what chances there were for a job. "Not good," said he, "but you can come along with me and try it; everything's as free as direct around a cow camp. I got three hundred head o' steers out here an' you can help drive 'em in."

In the morning, somewhat headachy and furry of mouth, we rode out a few miles to where two "breeds" were holding the restless herd, big, long-horned brutes, and taking over from them, started the long drive to the outfit. It was a glorious day. Prairie larks hung in the air above us. The golden plains stretched out before us rolling and boundless. At last I was punching cows! "Whether I get a job or not," thought I, "they can't take that fact away from me."

We pushed our cattle across the Big Muddy with some difficulty and camped that night near Manning's Ranch,[2] on the reservation, and were asked in to supper. His wife was a broad-faced squaw, silent and a good cook, as we later could testify. His sons were strapping youths, fleet of foot, as I found when after supper we staged sprint races in stocking feet. I lost by yards.

By morning light we were away, pushing our cattle north over plains now thick with ungrazed grass. Sometime in the afternoon we topped a long slope and saw below us the green trees and glint of Poplar Creek's water. The thirsty, lowing cattle, scenting it, started first trotting, then running, down the slope, and we had all we could do to stop them from plunging off the cut banks onto each others backs.

[2] Manning ranch headquarters were on the Big Muddy River fifteen miles northwest of Culbertson. The ranch was established by John Manning in 1896. He was born in Springfield, Illinois, on October 8, 1848. At the age of eighteen, Manning enlisted in the army at Kansas City, serving for the next eight years. In 1888, he moved to Vanderbuilt, South Dakota, with a caravan of five covered wagons. He later journeyed on to Culbertson. Manning died at Poplar, Montana, in 1918 (Mrs. Alfred Manning to Gene M. Gressley, January 11, 1965).

That night at dark we made Curran's, called a ranch by courtesy,[3] where a large corral gave us opportunity to dodge night guard. But the cattle were unwilling to be penned, and while chasing a runaway at gallop in the pitch dark my horse stepped into a badger hole and threw me sprawling. I lit on hands and knees and sprained a wrist badly (it still bothers me) but was so occupied in getting out of the way of the somersaulting horse that I didn't feel it. Later, however, when we unrolled our beds in a log shack, between my swelling wrist and mice which would persist in running over our faces, I got little sleep.

Another drive next day brought us over a long rise into view of the pastoral scene of the Bar Diamond outfit at graze on Poplar Creek. Down the slope in the distance were the white-topped tents and wagons, and here, there and everywhere the dots of grazing cattle guarded by a few mounted figures. We rode in, turned our bellowing herd into the main bunch, and soon were being made welcome by the "boys." I was introduced by Clint to the boss, with the words: "This here is Bob. He's lookin' for a job with the outfit. If you can give him one, do so, he helped me a lot with them cattle." Arnett was pleasant, read my letter from Brewster, said he would see what he could do on his return from Poplar agency in a few days, and meanwhile I was to make myself at home. I took him at his word, for just then the cook gave the traditional yell of "Come an' get it or I'll throw it away," and we were soon feasting on the hearty fare of a cow outfit on the range.

The next few days were spent in getting acquainted. I didn't have a regular job but made myself useful in any way I could. It was lucky I didn't have to work, as my wrist became increasingly painful, until it hurt even to ride at a walk, but finally got well with the help of a sling.

The cowboys were a fine lot, representative, I guess, of the better sort of range riders, with no taint of the ranch hand about them. Charley Bollinger, a "rannihan"[4] who came up with the trail herds from Arizona; Lew Gridley, a quiet southerner; Frank Bogart, called "Toler'ble," talkative and somewhat boastful, a Kentuckian;

[3] Livermore's observation may have been merited, as no one around Culbertson remembers Curran's ranch today.

[4] "Ranahan" usually connoted a top hand (Ramon Adams, *Cowboy Lingo* [Boston: Houghton Mifflin, 1936], p. 22).

Fox Cochran, the cook, who once owned his own wagon, but now was old, poor and pretty well shot to pieces, though still witty and amusing; Herman Hinz,[5] "the Dutchman," who was said to have killed a man in the East and who came west "for his health"; my friend Clint Otis; and the boss, Frank Arnett. The latter was a good cowman, though called, in the language of the range, a "shorthorn," that is, not an old-timer. He was minus a thumb, burned off by a rope, and was cursed with eczema, for which as a guard against the irritating sun he at times wore, with rather startling effect, a buckskin mask. Tom Reed and Alf Kingsley were the bronco twisters for the outfit, fine-looking chaps, but they left camp for the other herd shortly after I came, so I never knew them well.

All these men dressed in traditional plains style, white or checked shirts, loosely tied neckerchiefs, high-heeled boots, and spurs. They all wore gloves, for a rope fast to a plunging steer burns hands. "Chaps" were absent, never being worn in this grass country except in winter, notwithstanding movie "Westerns" to the contrary. Every man carried a revolver in a broad belt, variously called six gun, "cutter," or just a gun. Every saddle had a neatly coiled hemp ropes, sometimes called "twine" but never lasso, and rarely lariat. Hats were various, from the small black felts of the southerners to the wide Stetsons of the Montanans.

It seems queer now that a gun was a normal part of the plainsman's dress. It was, however. The country was still wild and the law not too evident. Only that summer, a party of Cheyennes had jumped the reservation and made east. Why, I never knew, but they made quite a stir before they were rounded up and returned by Uncle Sam's troopers. Cattle and horse rustlers were in the country, and only recently the cattle outfits had combined for a drive to clear them out. None of this I saw, but the conversation made the feeling of something about to happen ever present. After all, the last Indian uprising was only a few years before, and many people in the country knew and remembered Sitting Bull and his warriors, no

---

[5] The only one of this group of cowboys that the editor has been able to uncover any information about is Herman Hinz, who was born in Germany on May 11, 1872, and immigrated to the United States in 1900. During his lifetime Hinz was a jack of all trades; at one time or another he tried law, welding, ferrying, freighting, homesteading, carpentry, and coal mining. He died in Glasgow, Montana, on June 19, 1947 (Herbert Hinz to Gene M. Gressley, January 14, 1965).

more remote than the Great War of vivid memory is to us now. The main reason given me for wearing guns was the possibility of getting tangled in a stirrup and dragged, when there was nothing to do but shoot the horse; and though I never saw an accident of that kind, I can well imagine it might happen. Another was as a last resort, should the cattle stampede, to turn them by gunfire. They hated and feared it. The prevailing style was the single-action "Frontier model" colt. I had one, too, an investment of the previous year.

The cattle were mostly of the long-horned variety, and although I guess somewhat improved by breeding over the original Texas longhorns from which they sprung, they were far different than the tame whitefaces that make up the beef herds of today. Though getting fat and fairly handleable from good grass and easy treatment since the roundup, they were never dependable, and when on the move a stampede was always a possibility.

Arnett returned and, to my joy gave me the job of horse wrangler at thirty dollars a month. My duties were to guard the horse herd by day, bring them to the corral three times daily for change of mounts, drive them when on the move, and generally make myself useful. Among the useful duties was that of "snaking" firewood at the end of a rope from the scanty thickets in the arroyos leading to the creek. I was told to turn my pony into the herd, as no self-respecting cow outfit would make a man ride his own horse. I was allotted a string of my own, some seven horses, whose euphonious names in part were Roan Curley, Blue Dog, Troublesome, Windsplitter, Teeterlegs, and Billy Be Damned. The horse herd was later increased, until finally I had 120 head in my charge. The increase I well remember. A bunch trailed in from Oregon, good horses, but all with shoes or the remains of shoes on. As in this grass country our horses all went barefoot, it fell to my lot to pry them off. I learned a lot about taking shoes off animals who could kick with their hind or strike with their fore feet impartially.

Life passed pleasantly on Poplar. The object was simply to fatten two thousand head of steers as much as grass would do, until shipping time. For that purpose we allowed them to graze peacefully all day long, handling them as little as possible, and bedding them down at night in a close circle. One man, sometimes two, if the herd was restless, rode night guard around and around them, usually singing or whistling, both to amuse themselves and to soothe the cattle, for singularly enough a man's voice or whistle had that effect. A clap of thunder, or even a wolf's howl, was enough at times to send them all on their feet with one clash of horns, snorting and ready to go. At such times quick and careful work was essential. A stampede was a thing to be dreaded; it meant a wild gallop in the dark over a badger-hole-ridden plain to head them and get them to milling. Failure meant a scattered herd, to be gathered for miles around, some lost for good, all with many precious pounds wasted away. This, fortunately, never happened to the whole herd, though on occasion a part would take alarm and try to break away, when only hard riding and shouting brought them into the circle again.

Every few days we would move a few miles to fresher grass, when the herd would be pushed on first, to the new ground. All hands rolled their beds, threw them into the bed wagon and struck the cook tent. Cochran hitched his four-horse team and bounced over the roadless prairie in the rear. My job it was to dismantle the rope corral, gather my remuda and follow closely the cattle. On making new camp, the wagon was drawn up facing the cook tent, the corral ropes stretched in a diamond shape from its wheels, the saddle band corralled, and life resumed its accustomed routine.

For daily program, we were up by starlight, pulling on boots stiff with cold. In came the remuda driven at a gallop by the night wrangler, the "Dutchman." Breakfast over, each man roped his mount and rode off to his job. Though the corral was but a single rope, no horse—so well was his respect for a rope inculcated—tried to break through it. Horse roping was done by throwing the loop from the ground; it was bad form to twirl it overhead as in roping cattle. All of us practiced continually on every convenient object, and I became fairly expert at it, but never like the old hands who could throw a three-foot circle the length of their thirty-foot rope neatly over their victim's head, with rarely a miss.

At supper time, when the cattle and horses were well fed and quiet, we gathered at camp for a loaf before supper and the night activities. Poplar Creek was a clear, quiet, sandy stream, forming deep, broad pools at intervals. In these we would ride our horses, both man and horse stripped of clothes and saddle, and swim them as much to their enjoyment as ours. I was a better diver and swimmer than the plainsmen—in fact, most of them couldn't swim at all—and astonished them by bringing up the white pebbles they would point out from the bottom. They couldn't believe that I actually opened my eyes under water, until I challenged them to throw in a silver dollar, which being retrieved, convinced them. After that they all tried it and were as pleased with their accomplishment as a child with a new toy. The dollar was lost eventually, and the place received a name, "Lost Dollar Camp."

The only drawback to our pleasure was the curse of mosquitoes, which arose in clouds each evening, to cover horses and cows and drive them frantic. Ourselves, what with fire smoke and clothes, we could keep fairly comfortable, but we could do little for our charges, until the idea of building a smudge for the horses occurred to us. For lack of wood, we gathered piles of dry cow chips—now and in the days of buffalo the only fuel on the plains—and toward evening touched them off. The horses, intelligent animals, soon caught on, and would even come to camp voluntarily and wait patiently for the smudge to be lighted. We called it their evening smoke. It was comical to see them jockey for places nearest the piles, and when the smoke started, edge into it, until tears ran down their noses.

Arnett took one of the wagons and, with Alf, Tom and Clint, with their saddle strings, pulled out for the other herd on the Big Muddy, leaving five of us to guard the beef herd.

The Indians were on the move to their summer hunting grounds. Their trail lay past our camp, and we heard in some way that they were coming, and rode to meet and look them over. First came an outrider dressed in a soldier's blue uniform, but with moccasins and a feather stuck in his broad hat. We looked each other over after a greeting "How," and after ineffectual attempts to talk by signs and pigeon English, rode on to the main outfit. This was a motley-clad crew of men, squaws and papooses, some horseback, most in two-wheeled "Red River" carts,[6] accompanied by a barking pack of curs who snapped at our heels as we dignifiedly rode the length of the procession, then back on the other side, with no word exchanged. I remember vividly a lass of twelve or so sitting beside a dark squaw,

[6] The Red River carts derived their name from the Red River Valley in Minnesota and Manitoba, where they were utilized in the fur trade. The cart was far from an imposing vehicle; a square box set between huge wheels, it was constructed entirely of wood and leather. "Shaggy ponies or ponderous oxen" were its motive power. For a very realistic description of the Red River cart, see "The Red River Trail," *Harper's Magazine*, XVIII (April, 1859), p. 615.

who herself was red-haired and fair as we. I have often wondered
how she came there—a throwback to some captive ancestor, or one
of the post whites on tour with her Indian friends.

The next day another section of the tribe came by. This crowd,
perhaps poor relations, traveled by travois, long poles lashed to the
pony's back, with a canvas holdall slung between, in which were
carried tents and camp belongings. A couple of brawny bucks, their
hair in two braids, stalked into our tent, Indian fashion without a
by-your-leave, while we were eating. With a grunted "How," they
squatted and watched us with beady eyes. Presently, in came a little
old Indian bent with age, his face a mass of wrinkles, his garb hidden
by a long linen duster such as stage drivers wore. Without a moment's
loss of time, he pulled out his long square-stemmed pipe and filled
it from the ample buckskin pouch which hung from his waist, and
puffed complacently to the end of our meal. Then he spoke sharply
to the young bucks, who with one accord rose and vanished as
abruptly as they had come. After a beaming, all-embracing smile,
he then spoke as follows: "Me big chief!"—pause, with no comment
from us. "Me got three squaw, me got papoose, me got "wokapominy" cow, me got tea, me no got shookoo!" Still no comment
from us, and a repetition of the lack of "shookoo" from the chief.
"He means sugar," said I, inspired, and the chief's grin widened.
"Goo boy," said he, and rose and shook my hand. So we gave him a
sack of sugar and watched him disappear over the divide toward the
dust of the departing travois, his feet belaboring his horse's sides,
his duster flapping in the breeze.

Arnett came back and chose me to go with him with a beef issue to
Poplar agency. We rounded up the herd and cut out 150 head, or
rather the others did,—I had to hold the "cut." It was the first time
I had seen cutting out, and I watched to good effect for later experience the way the rider would single out his steer and, once picked,
leave it to his pony, who never lost that particular animal but
followed his every turn, until once on the outside, a wild race
ensued, the steer to get back to the obscurity of numbers, the pony
to head him. The pony never lost the race.

The "beef issue" was the reason of our being on the reservation.
A polite fiction was maintained that we were fattening cattle for the
Indians, but so far as I know, only two or three hundred were driven
in to them. The rest lined their ribs with virgin grass for the Chicago

market. Advantage was taken of these issues to discard as many defective cows as Arnett thought "wokapominy" (the government) would stand for. There was a goodly percentage of "lump jaws" and aged steers who remained gaunt, no matter how good the feed. I think Frank felt a little guilty, as when we had cut complete he asked me if I didn't think they were a pretty fair lot after all, but truth prevented my easing his conscience. Those lump jaws and bony hips were far too prevalent.

We eased down the creek and camped during the heat of the day, then resumed our drive by night. There was a fascination in the night drive; the sharp line of the prairie horizon seen by the light of stars so brilliant they seemed fairly to crackle; the only sounds the swish-swish of the cattle moving through the long grass, and Arnett's whistle from across the backs of the herd. After midnight, we camped again and took turns at sleeping and riding night guard; then, after a reviving coffee, drove through to the Indian corrals at Poplar, and threw our cattle on water and feed while we ate and rested. We trusted them too far. They didn't like that place, with reason, and of a sudden raised their heads, gave a snort and stampeded. Two miles of hard riding, shouting and cursing it took to round them up and return. Then the Indians arrived, and together we held and subdued them.

The post was composed of the government houses, a few log shacks, a store, and the receiving pens for the beef. All around were the tepees of the Indians with their owners gathered for the issue, clothed in every variety of dress from blankets and buckskin to some combination of white man's discards. I saw one buck with a once white "boiled" shirt, tails flowing free, and a red sun painted on the bosom. Apparently he thought it an opportune spot for decoration.

I was left in charge of the cattle, together with a young Indian with long black hair. We rode around the herd in opposite directions, passing each other with the customary "How." At length, he stopped and asked me for the "makin's," and as he rolled his cigarette said: "You from East?" "Yes," said I. "You go to college?" "Yes." "You play foo-ball?" "Yes." "Me too, I play Carlisle!"[7] It developed that he had played against Harvard the year before, but

[7] A famous school for Indians at Carlisle, Pennsylvania. Jim Thorpe was one of Carlisle's best-known alumni.

as I was not called on in that game, we didn't play against each other. Still, we were both on Soldier's Field at once. Such is coincidence.[8]

The cattle were driven onto scales, then into a vice, the U.S. brand stamped on them, and turned over to the Indians. Some slaughtered them at once, and I guess had a bit of sport chasing them à la buffalo before doing so; but Arnett was in a hurry to return, so we gathered our saddle and pack horses and left the lively scene.

Arnett liked a song, and I exhausted my repertoire on the trip. Perhaps by that medium I unconsciously established myself in his good graces, for on return he told me that I was to accompany Clint Otis on a "repping" tour to the French-Canadian roundup in Assiniboia.[9] This was a real promotion, from wrangler to work with cattle, and aside from the novelty of the trip, I was thrilled to be chosen.

We moved camp a couple of times, and in anticipation of my absence I was kept busy snaking in wood until I had every snag in riding distance piled up. Events my diary tell of were ever increasing plagues of mosquitoes; a hard fall taken at gallop when a badger hole sent my horse headlong, and in falling caught my spur in the web of the cinch—had not my spur strap broken, I would have broken a leg at least; large numbers of ducks gathering on the lagoons of Poplar, making me long for a shotgun; a band of wild horses, led by a fine-looking stallion who tried to cut out some of my remuda and was sent flying by a shot or two.

The day came for departure. Clint and I got together our string of six good pack and saddle horses and headed north. As we neared the Canadian border the country seemed wilder somehow. The grass

[8] Soldier's Field is an area of land on the south side of the Charles River, which Henry Lee Higginson presented to his alma mater, Harvard University, in 1890. Referring to it as a "playground," Higginson wrote President Eliot on June 5, 1890: "The gift is absolutely without condition of any kind. The only other wish on my part is, that the ground shall be called The Soldier's Field—and marked with a stone bearing the names of some dear friends, alumni of the University, and noble gentlemen, who gave freely and eagerly all that they had or hoped for, to their country and to their fellow men in the hour of great need—the war of 1861–1865, in defence of the Republic" (Bliss Perry, *Life and Letters of Henry Lee Higginson* [Boston: Atlantic Monthly Press, 1921], p. 330).

[9] There are two regions in Canada with the name of Assiniboia. One is between Saskatchewan and Alberta, the other fifty miles west of Winnipeg; Livermore is obviously refering to the latter.

was more luxuriant and flowers bloomed in profusion; bones of buffalo were seen, sometimes in groups, lying just where they had been slaughtered, some so recent that the horns were hardly weathered; a couple of antelope gave us a glimpse of their white rumps as they fled.

Lost Lake,[10] over the border, was our destination for the night, and toward evening we found it, a testimonial to Clint's bump of locality, for it was only a dot of grassy water surrounded by boundless identical rolling plains. We rode in and made camp, disturbing a flock of ducks and a cloud of mosquitoes. A fire dispersed the latter a bit, but we ate our frying-pan supper in a pouring rain and thunderstorm, and after picketing our saddle horses took what comfort we could by rolling in our blankets and pulling the tarpaulin over our heads. But little sleep were we to have that night; the storm increased in violence until the thunder was one constant roar and the lightning so incessant we could see by it our loose horses drifting away down wind. I jumped on my staked horse bareback and rode after the bank, drenched and somewhat scared. I felt a target for the artillery of heaven, which seemed to send bolt after bolt in search of me, alone in that bare plain, with almost personal animosity. We found the horses, though, drove them back, hobbled them and crawled back in our soaked bed to get what sleep we could.

The morning sun dried our soaked clothes and bedding as we rode, and the troubles of the night were soon forgotten. Greener grass and patches of small timber in the rougher country we were entering showed us we were leaving the plains with which we were familiar, and soon a faint trail led us up a creek through hills to Bonneau's Ranch.

The family, consisting of *père et mère*, Pasquale, the oldest son, the "boys," and last but not least, "Bina," the pretty daughter, made us welcome in true range style, and we reveled in the luxury of a roof and a sit-down meal. It was there I first encountered that curiosity, a "jumping Frenchman." The second son, a strapping youth, was carrying a pail of milk from the springhouse, when Pasquale, the mischievous one, called "Drop it!" Upon that the "jumper" gave a loud yell, hurled his pail into the air and went to sparring vigorously at nothing. Again at supper Pasquale cried

---

[10] About thirty-five miles northwest of Whitetail, Montana.

"Hit him!" and the boy struck out at the nearest object, which happened to be his brother's head. Clint and I struggled between politeness and our sense of the ludicrous, but finally joined the family (all except the unfortunate one) in a roar of laughter. It is a form of nervous disease peculiar to French Canadians of which I had heard but had never seen.

The wolves had been very troublesome around here; Bonneau told us that several full-grown steers had been pulled down by a single giant wolf, and showed us the carcass of one. The wolf catches the animal by the nose and throws him, then gets a throat hold, and it is all over. While out wrangling horses, unarmed and riding bare-back, I saw one, not the giant, however, trotting ahead of me on a trail, carrying a big chunk of meat. I spurred up, swinging my rope, but he hurried his pace not at all, just gave me a backward look and showed a glint of teeth. My horse didn't seem anxious to near him, and I thought, after all, a lively wolf at the end of a rope, with no saddle to tie to, isn't so hot, and left him alone.

We rounded up Bonneau's herd and cut out about forty head of Diamond cattle, and a wild, snaky lot they were. Once separated from the herd, they set out on a dead run through that broken country, apparently determined to lose themselves in the wilds of Canada, with both of us after them in peril of our necks. Clint finally headed them and left them for me to hold while he got lunch at the ranch, then relieved me while I did the same. He was supposed to push on south and I to catch up with him, but while I was saying goodbye to Bina, back he came on the gallop, saying he had forgot to get some baking powder, and for me to follow up the herd. I guess a last look at Bina had more importance than either baking powder or cattle.

I found them on a trot, wild-eyed and snaky as ever, but fortunately headed right, and soon got them strung out properly, and, joined by Clint, drove steadily south until night. I took the first guard that night, and it was nearly the only one ridden, for the cattle stayed restless and ready to break, and carried me far away from camp in the pitch-black night. I had all I could do, whistling, singing, and shouting, all the time riding around and around them, to keep them from stampeding. By the time I had them lying down I had no idea where in the world I was, and as to waking Clint for relief, I didn't dare to leave my charge for a minute. The cattle finally solved the situation, just before dawn, by carrying me, in one of their numerous attempts to make off, within sight of the dimly seen forms of grazing horses, and I knew I had made a many-mile circle back to camp. I got an hour or two of broken sleep in a rain which didn't prevent mosquitoes from tormenting, while Clint took his shortened turn. We were both glad to be on our way by morning light, breakfastless, to a more congenial spot.

As usual the morning sun warmed us, and we found respite from the night guard on open prairie by arrival at a little ranch, a small log hut with a sod roof, and a fenced pasture, into which we turned our stock. The owner, one Thompson, was away, but, range fashion, we made ourselves at home, cooked ourselves a meal and even enjoyed the luxury of milk from a ranch cow which we roped and milked inexpertly into a tin can. Bina's parting gift of baking powder had spilled and spoiled in the pack, at which Clint was mightily wroth but cooked a "dough god" anyway, which, when extracted from the oven, had risen beautifully. It seems old Fox had mixed the flour in advance, which accounted for our having no baking powder in stock.

Next day, man and beast rested, we pushed on to a noon camp at the great flats where Eagle Creek joins the Big Muddy.[11] Here, I saw thousands of curlew which I took to be of the sicklebill species, and again I wished for a shotgun. Then, on down Muddy through miles of luxuriant bluestem grass shoulder high to the horses; and hard it was to drive the cattle away from such succulent feed. In mid-afternoon we crossed a rise from which we could see far distant and below on Wolf Creek,[12] to which it had moved in our absence, the Bar Diamond outfit, wagons, cattle and horses, like miniatures in the clear prairie air, but so deceptive was appearance that it was nearly dark before we pulled in and turned our consignment loose in the big herd.

We found the storm which beset us at Lost Lake had played havoc with the home camp, blowing one tent away and ripping the other to pieces, so all hands were camping under the sky. The outfit was on the move, slowly drifting eastward toward the shipping pens near

[11] Eagle Creek flows into the Big Muddy fifteen miles above Culbertson, Montana. The Big Muddy confluence with the Missouri takes place at Culbertson.

[12] Wolf Creek joins the Missouri at Wolf Point, Montana.

Williston, but so as not to lose any precious weight, only a few miles a day. Sieben, one of the owners, had joined us and was busy with his shotgun, so that we had duck for every meal that we wanted them. Fox Cochran's inflexible rule was that every man should dress his own game, and it was there that I learned how to pick a duck properly, cleaning up as you go, and leaving no feathers behind.

My status as a result of the successful "rep" trip had improved. I was tacitly promoted to cow hand as well as wrangler, and thenceforth rode regular night guard on the cattle. I loved that work; when wakened for my guard, the sleep once rubbed out of my eyes, to saddle my picketed night horse, ride until the dark mass of bedded cattle came in view; then, taking over from my predecessor, to ride slowly around them, not too near, always giving warning of approach by a low whistle or a scrap of song; while the bright stars made shadows on the grass, and the Big Dipper slowly turned and marked the hour;—that was fun. It reminded me of my trick at the wheel at sea,—the same feeling of responsibility, the same time to be alone with one's thoughts.

I got on well with the cowboys; they were gentlemen by instinct, whatever their origin, and when they had once sized me up and saw that I was anxious to learn, were more than considerate and friendly. "Tolerable" alone was somewhat "or'nery;" he picked a good deal on me at first, and we once came near explosion when I became exasperated at some semi-threat he made, and told him to "try it on if he liked." I remember there was dead silence in the circle around the camp fire for a moment. Threats were serious in a gun-toting country. Then, Bollinger said in his slow way: "Tol'able, yo' better think twice before yo' tackle the kid; I was a-noticin' of him stripped the other day;" and Tolerable thereafter kept his peace. Again, when later I felt that I was one of them, and announced that I was too old to be called "kid," one and all thenceforth they used my name or some friendly nickname like "hackamore" (a rawhide halter), partly from my liking for using one instead of a bridle, and partly in parody of my last name. Gridley had a quaint humor and a southern drawl. One day when the cattle he was guarding persisted in over-running my horses' range, making me ride hard to keep them separate, he came up to me as I fumed and sweated, and said: "Was my bovines a-hookin' your equines, Bob?"—and a laugh and his quiet assistance restored good humor and straightened out the tangle.

The conversations around the fire were of cattle, the market for beef, (I remember it was eighteen cents a pound that summer, an all high) the latest doings in Culbertson. One incident there was of an Indian from the reservation, who rode in, announced that "his heart was bad" and started smoking things up. They shipped him back in an empty cattle car, dead clay. Talk was of taking up land and settling down. Most said they would never want to do anything but ride the range, but Arnett said: "You boys better take notice of what's goin' on; the big outfits' day is over; ten years from now this country will be all-fenced up." And he was right, except he allowed too much time. A few days later we came on new fences along the Little Muddy,[13] where none had been when the roundup started.

Local politics had little interest for the men. There was as yet little competition in that unorganized land; but they took a lively interest in world affairs, a smattering of which they would get from the very occasional Chicago weekly which found its way to camp. The Dreyfus trial was on at the time, and many were the arguments as to his guilt or innocence. An old ranchman whom I later met used to draw me out on opinions, and then shatter them with his own shrewd argument. He referred to General Mercier, one of the protagonists, as "this yere General Meredith." Once, in a discussion of some question between England and the United States, one of the boys referred to me, saying, much to my surprise, "You're English, ain't you Bob?" which I indignantly denied; and the Dutchman confirmed me, saying: "Hell, no, he ain't English; I've heard that kind of talk; it comes from 'round Boston." My broad *a* had evidently not been sufficiently subdued as yet.

As to adventure, I am afraid my tale is a tame one. Risks there were, of course, usual to a life with horses and cattle of uncertain temper. Perhaps the greatest was that of falls from hidden badger holes, of which I took more than one, as already recorded. One risk I ran unconsciously, which might have been the most serious of all: When on Poplar, one of the night horses, driven frantic by mosquitoes, broke loose, and I tumbled out of bed, and, just as I was, jumped on my own horse bareback, rode after the runaway and caught him just as he joined the main bunch. The "night hawk,"[14]

[13] The Little Muddy empties into the Garrison Reservoir near Williston, North Dakota.

[14] The "night hawk," the night herder, was one of the "fellows who claimed he swapped his bed for a lantern" (Adams, *Cowboy Lingo*, p. 25).

Herman, rode up in the dark. "Where are you going with that horse?" said he. "Back to camp," said I, surprised. "Yes? Where are you camped?" I saw then that he didn't recognize me, and explained. He later told me that, seeing a figure without a hat, and shirttails flowing, he had taken me for an Indian skulker, had me covered, and was deliberating whether to shoot first and inquire afterward, but, fortunately for me, decided on questioning.

The horses were well "gentled" for range stock, and there were no outlaws among them, though one or two had something of a reputation. When first saddled in the chilly morning, they were apt to "hump their backs" a bit. Bollinger's favorite, named "White Man," invariably gave an exhibition, but it was all understood between them. Once, Arnett's horse, too quickly spurred after mounting, went after him and, as the boys put it, was "sure lookin' for man," bowing, his head buried between his front legs, and each jump getting higher. Arnett, to do him credit, sat him well and rode out the storm, though he had to "pull leather." My string was well behaved on the whole, though "Billy Be Damned" gave me a ride or two, and my own horse, "D.K.," rarely ridden, performed his circus trick of hurling himself backward once or twice. On the whole, my riding stood comparison well with the others.

Among the cattle were some who had caught mud fever, a disease which swelled their feet and legs to elephant size.[15] The rough surgery practiced was first to rope and throw the patient. Then, two men would saw a lariat between the cleft of their feet, until the swelling burst and the pus drained out. Then they swabbed out the wound with raw turpentine; and while one rider set tight on his rope to hold the steer to earth, the others would scramble for their horses. The holder would shake loose his rope, and a maddened steer would rise and charge the nearest object. Sometimes a horse would object to being mounted, or else the holder would slyly turn loose too soon, and a good imitation of a bull fight would ensue, the dismounted one the object of the steer's wrath, the riders whooping and riding him off, and all having a good time except the steer and the man afoot.

[15] According to Dr. Paul Stratton, Professor of Animal Science, College of Agriculture, University of Wyoming, Livermore is probably referring to foot rot disease, the symptoms of which are swollen hoofs and lameness; in advanced stages the malady is sometimes complicated by an arthritic condition (O. H. Seigmund, [ed.], *Merck Veterinary Manual* [Rahway, New Jersey; privately printed, 1961], p. 1152).

One morning the cattle moved out ahead as usual, and through long delay in breaking camp my horse herd had scattered. Gathering them took time and left me far behind. Cattle, riders, and wagon had disappeared, so I took their general direction. An hour's drive brought no sight of them, and, getting in front of my herd, I could find no tracks. There I was, one hundred head of horses, hungry and hard to drive, on my hands, on plains whose every swell looked like another, and only knowledge that the course was vaguely eastward, for my guide. I was worried, no mistake, and had visions of wandering for days, losing my horses far and wide, but finally took courage and laid my own course, with the result that in the afternoon I came on the outfit camped on Painted Wood Creek[16]—a most welcome sight. I tried to look casual as I rode in, but silent grins and subsequently some jocular remarks about "the wrangler must have had some visitin' to do," etc., kept my crest down for days.

We drifted along the next few days, well settled in the routine of moving. The line camp, where Alf and Tom held the "dogie" herd, reached, we spent two days cutting out those of our beef herd which were not to be shipped and gathering accumulated strays. I was mounted on a good "cut" horse and given my first try at cutting. My observations of the art on Poplar came in useful, but I was hard put to it to sit the horse, who with no guiding rein from me twisted and turned in the midst of the herd, more agile than the steer he followed.

The country was still untouched by civilization. Wild horses, a big stallion with flowing mane and tail in the lead, fled like antelopes at the sight of us. Numerous small lakes dotted the plains, each filled with ducks and shore birds, now gathering for the fall migration, from which Sieben drew ample toll.

Night guard was now the absorbing feature of the drive. It was all important that the cattle should be kept quiet and not lose any of that hard-won poundage. I have later wondered that they entrusted to me, considering my inexperience, my share of that duty. I overheard Bollinger and the "Boss" discussing whether I should go on, the last night before shipping, when Arnett said, "We can't take chances now." Charley replied: "He's real gentle with 'em." Perhaps that was the reason. I did get on well with them, for a fact. Whether or not they liked my singing, they at least didn't mind it, and I

[16] A small tributary emptying into the Missouri River near Bainville, Montana.

practiced on them a lot. I hear so-called cowboy songs now which I doubt were ever known on the range. Our boys had no connected songs, just inarticulate snatches of minor-key ditties, without beginning or end. The nearest to a complete range song which I sang went this way, in a rather monotonous minor strain:

> Now cattle, cattle, cattle, I'm callin'
> Stop your pushin', millin', crowdin and a-bawlin'
> Lie yo' down and sleep
> While my guard I keep
> And watch the weary hours of night a-crawlin'.

At length we reached the fringes of settlement on the Little Muddy north of Williston, where fences and haystacks proclaimed the end of the range.

From then on to the shipping point on the railroad we skirted the fences and held our herd on the sparse grass with difficulty. I record my last night guard as a tough one. The cattle kept me continuously on the move, whistling, shouting, and singing as I rode. One minute they would be quietly bedded, then, with a clash of horns, they would rise as one and stand ready to break. In spite of my efforts, about twenty broke away in the darkness, and to hold the main herd I had to let them go, but fortunately they came back when the others, praise be, didn't follow. A very relieved young cowboy turned his charges over to Gridley at the end of his guard, with the remark "They're bad tonight, Lew." Horses also gave us trouble; crazy for green grass, they broke into hay fields, and, unless watched every minute, tried to take the back track to remembered richer feeding grounds. Busy with my double duties, I lost a part of the band several times, much to Arnett's disgust, and had to make forced gallops of miles to head them.

Shipping took us two days. The bands were separated and driven down to the pens, then rushed aboard the cars in groups of a dozen. There was no gentle handling of cattle or sparing of men and horses in this work. Two thousand head must be gotten aboard in jig time. Here it was that "Billy Be Damned" lived up to his name and got me a small reputation as a rider. In a hurry to ride to the lone saloon for some thirst-quenching steam beer for the boys, I spurred him too vigourously and he "went for man." It was a good deal like a whirlwind, but I sat him by good fortune. Needless to say, on return with the bottles I rode very gingerly.

Shipping over, feeling strangely free of the slow-moving cattle,we pushed our saddle band on for a couple of camps to the old Diamond Ranch, a cluster of log huts and corrals above Culbertson, and rode into town for an evening of enjoyment of that town's very limited merrymaking facilities. The next day I got my pay and bade the Diamond outfit a reluctant good-bye.

I now found myself a cowpuncher out of a job. My entry to Culbertson wasn't too auspicious. I sat down at the common table of the one small eating house, amid a mixed group of townspeople and travelling men. In front of me was a bowl of soup which I found surprisingly rich and good; but as I ate, became conscious of a lull in the conversation and the gaze of all upon me. I was then aware of the fact that I had eaten the gravy for the whole table! Either politeness or my rather wild range appearance restrained comment, but my face burned at the realization of misdeed.

I thought of travelling a bit and seeing the country, but I was taken really ill with a digestive upset and spent several nightmare days at the one hotel, an establishment thrown together with boards, its only plumbing a bucket of water outside the door, into which everyone dipped for a wash or drink. I found myself at the end pretty weak and pale, with money alarmingly low. I had to get a job, and would have taken one even with the hated sheep if there had been any nearby. I was getting very down in the mouth when one Archer, a ranchman, I suspect at Arnett's behest, offered me a job at haying. He expressed considerable doubt that I was husky enough to stand the work, and I guess I did look washed out. My persuasion won the day, however, and I turned my horse in pasture, drove with him the few miles to his ranch, changed boots and spurs for moc- casins, and went to work.

One Russell Mann, a robust ranch youth, and I each had a two- horse team and rig of which we had sole charge, and with which each were to haul four loads a day of the wild bluestem hay which Archer mowed ahead of us. This hay must be pitched from the ground to the rig, then from the load to the stack, singlehanded. There is a knack about haying, of which at first I knew nothing. I lifted the heavy forkfuls by main strength instead of by leverage, and for the first few days I suppose no youth was ever lamer or more exhausted. The

Maine paving stones were nothing to it. I had no bed, simply unrolled my blankets on the floor, but it made no difference; that was
the softest couch I ever knew. Mann was friendly and showed me a
trick or two, which, added to my determination to prove Archer
wrong about my staying ability, enabled me to weather the gale and
haul my four loads each day. I did, however, strain my back so that
it handicapped me for weeks afterwards.

Archer's ranch was a simple affair of logs within a few rods of the
cut bank of the Missouri, which flowed by in a boiling, muddy flood.
From it we drew our water supply, which, when settled in a barrel,
provided a cloudy but drinkable fluid. Archer said he preferred the
flavor to that of spring water, and it *did* have "body!" Our main-
stays of food were bacon and potatoes, varied by an occasional
catfish caught on night lines set on the river. We all tried our hand at
cooking.

The plague of the spot was rattlesnakes. Where west of the Muddy
we had mosquitoes but no snakes, east of it there were both. The
snakes liked to get under the new-mown hay. More than once I
picked up a forkful, to see a buzzing reptile slide out of it. Some
even rode in with us and were pitched onto the stack, where we
could hear them rattle as we trod it down. Once I killed a huge one on
our very doorstep. Familiarity bred contempt,—we simply listened
for the rattle, and, once located, either killed the owner or ignored
him as long as the sound was not too close. Antivenoms were
unknown to us, or cared for, in those days.

Smoke from great prairie fires filled the air by day, and their glare
lighted the horizon at night. This was a serious matter in a cattle
country dependent on range for winter forage, and the call for fire
fighters went far and wide. We saddled up and galloped miles toward
the fire, but it had swept too far for us; only the blackened plain
stretching north as far as we could see met our eyes. The method of
fighting fire was to wait till the wind had died down at night, then to
kill and split a cow from end to end. Two riders would tie onto a
foreleg apiece, and drag the carcass, bloody side down, along the line
of flame, while following beaters would extinguish the embers.
Crude, but effective.

I earned my pay at Archer's but made the usual blunders due to
inexperience. Once I trusted my team too far by leaving them untied,
while stacking. Half-broken, they bolted, and disaster would have

followed had not a wheel caught in a gate post and delayed them long enough for me to spring to their heads and drag them to a stop. About all the damage was a broken post and a strap or two, but Archer made much moan.

The stacks were full, and haying over. Archer's parting words on paying me off were: "Bob, you're nice to talk to an' have around, but ye've got a lot to learn. Trouble is, you ain't had to scrabble to get a little outfit together like I have, an' then you go an' bust 'em up. It's aggravatin'. I will say, though," he added, "Yo' worked hard an' yo' did go after them runaways good." So we parted friends, and I took the lesson to heart.

My wages and a timely fifty dollars from home put me in fair funds, though not enough for car fare home, so I decided to ride as far as I could and trust to luck. I went to Culbertson, wheedled a pack pony out of Arnett for a few dollars, bought a tarpaulin, a frying pan and a little grub, broke the "tarp" over the pack pony and lashed it on with a squaw hitch, saddled D.K., and headed east. My back was very sore; I had evidently strained some ligament severely, and for a day or two I could hardly mount the pony, or, once in the saddle, ride out of a walk. A couple of days of slow riding and lonely night camps on the prairie brought me to Williston. At one of these camps a settler, seeing my fire, came over and invited me to his cabin. It was a simple affair set up on the raw prairie, with no sign of cultivation as yet, but pride of possession shone in his face as he ushered me inside and introduced me to his family. Their greatest treasure was a scratchy gramophone which ground out Sousa marches, to their never ending joy and wonderment. I too recorded my appreciation of the march of science at hearing music from a full band in such a remote wilderness. Someone has said lately of the radio that it has banished loneliness, and the phonograph had somewhat the same effect in its day.

I intended to rest up at Williston, but a drayman offered me a job at unloading lumber from a freight car, and I flew at it. Handling boards all day, far from making my back worse, seemed to cure it, to such effect that I rode out of Williston next day practically sound again. Williston, even in the short time I had last seen it, showed evidence of the advancing tide of immigration in its new buildings and change of garb from boots and Stetson to straw hats of farmers, and even a white collar or two. Here, I struck up an acquaintance with two young men driving a wagon east to the wheat harvest, threw my pack on their wagon, and thenceforth travelled in their welcome company. We stopped that night near a camp of half-breed Indians, men, women and children, trailing, they said, to Bottineau, near the line, a shaggy group, more red than white. From their midst a white man wandered over to our campfire and evinced a desire to join us, so we added him to our family. He was a poor specimen, without even a blanket of his own. We thought we could make use of him, but when he was delegated to cook a duck and found to have put it in the kettle to boil with all its insides and most of its feathers, Indian fashion, we regarded him henceforth as a passenger only.

Bill Moore and Frank Durand were decent chaps, superior type of harvest hands, who, bent on bettering themselves, probably later became prosperous citizens of whatever spot they chose to settle in. They argued with me frequently to give up my wild way of life and join them in the wheat field to make a stake, but unsuccessfully, though I did not tell them the reason, that I was due in Harvard. Bill said at the end of the argument: "Well, Bob, I suppose you won't never be nothin' but a cowpuncher, thinkin' about nothin' but ridin' an' cattle, when you might get you a good piece o' land and settle down."

For some days we travelled over plains, still largely unsettled, camping at night by one of the numerous reedy lakes that dotted the prairie. Ducks swarmed in these, and as our arms were a rifle, a pistol, and an antiquated muzzle-loading shotgun which usually failed to go off, our bag was small. Our usual method was to shoot with a rifle or pistol into the midst of a flock on the water, then I would strip and swim with my horse to retrieve the occasional victim. It is the first time I ever heard of using a horse as a retriever.

Ranches and hayfields got thicker, and at last we were riding on a road between fences. Ten miles from Minot, North Dakota,[17] a ranchman offered me thirty dollars for my two horses. I accepted, threw my saddle in the wagon and mounted D.K. bareback to ride to the little town where delivery was to be made, only to be promptly bucked off, a parting gesture from my faithful steed. Luckily, the rancher didn't witness the episode, and payment was duly made.

[17] In 1900 Minot had a population of 1,277.

The wagon having gone on, I had to walk the ten miles to Minot, no easy job in high heels. It was a lonely walk, howled in by the coyotes, who, safe in the darkness, slunk alongside, intrigued by the unusual sight of a man on foot. Long after midnight I reached town, and found with difficulty lodging in a miserable insect-ridden rooming house. Next morning I retrieved my bedroll and saddle, said good-bye, and boarded the train.

I had estimated that my finances would just get me home, by paying the cheaper fare, via the Great Lakes from Duluth to Buffalo; and after running the gantlet of hotel runners urging that "all the Western boys come to our place," boarded the "Northland" for the passage. I would have enjoyed the novelty of the trip, the comfortable ship, Superior, so like and yet unlike the sea, with its sparkling blue water and land-free horizon, the green shores of the straits, much more had I not found that my ticket didn't include meals, and for three days I subsisted on the lightest of fare in spite of the healthiest of appetites. To aggravate my fix, there was a returning Alaskan gold miner simply dripping nuggets, who took a fancy to me and insisted on treating me to cocktails, which only stimulated my appetite the more. The free lunch on the bar helped a little, but I was too proud to suggest that my rich acquaintance include an invitation to dinner. I'm sure he was surprised that an apparently normal young man skipped so many meals. I was stoney-broke and hungry homecomer when at last I gained the familiar and hospitable doors of 34 Alveston Street, Jamaica Plain.[18]

Though my cowpunching expedition offered no stirring adventures and had rather a tame ending, I have always been glad that I rode a little while with real riders of the range and saw something of the life of an old-time cow outfit, ranging its cattle on free and open plains now gone forever. Last year, I looked out of the car window near Culbertson, to see wire fences, dreary ranch houses, and meager crops. The very tules had been grazed from the dry pond holes where once we camped by lush grass, and migrating ducks splashed in.

[18] Once a suburb of Boston, Jamaica Plain now is in southwest Boston, bordering on Brookline.

V I

College Years

This was my senior year in college. I was often told that if I only knew it I was having the happiest time of my life. Though I don't agree entirely, my four years at Harvard were reasonably joyous, and certainly in comparison with later life, carefree. Looking backward, I am of course conscious of many errors and lost opportunities, but I don't know that I regret them much. My time at Harvard was in the height of Eliot's elective system, when a man could choose any course he wanted so long as the requisite hours were filled.[1] I imagine it was harder to get in then but far easier to stay in.

Although I entered with honors, I soon was infected with the easygoing ways of most of my acquaintances, and thenceforth contented myself with marks which kept me comfortably out of the Dean's office. I think I only flunked an examination once and only got one mark higher than "C" in all my college career. I would have pursued "snap" courses even more than I did had not father, casting an eye on my freshman schedule, and spotting an absence of Latin, Greek and math, laid down the law that I was to take those subjects for at least two years, and I revised the list accordingly. I'm glad he made me do so.

Through sheer laziness I missed many of the contacts with older men that some of my friends enjoyed, as with "Copey," Shaler,

---

[1] Although the elective system at Harvard dated from 1825, President Charles W. Eliot broadened it. "With the extension of the elective system to the Freshman year, in 1883–1884, the President announced the 'practical completion of a development begun sixty years ago'" (Samuel E. Morison [ed.], *The Development of Harvard University, Since the Inauguration of President Eliot, 1869–1929* [Cambridge: Harvard University Press, 1930], p. xiii).

Dean Briggs, Wendell, and others.[2] They were just teachers to me, to satisfy in class and go my way. I found my pleasure in athletics and outdoors in general, and in growing social acquaintanceship. I lost some of my backwoods habits and joined a raft of clubs from the Institute and Dickey to the final A.D., all of which kept me delightfully broke and made me lasting friendships. Although there is plenty to criticize in the club system in Harvard, so far as I was personally concerned, it suited me to a " T." Secret societies were fortunately largely missing, but nevertheless the non-visiting rule and the close bonds of companionship and good fellowship cultivated within the walls of each club tended to narrow one's acquaintance and thereby make one miss knowing well a lot of worth-while men who went to the other clubs or made none. I value tremendously the enduring friendships I have made and count the good times I have had in college and at reunions among my pleasantest contacts; but I have never been one of those who, having made the club of his choice, believed that his chief object in life was attained, that he didn't need

[2] Charles T. Copeland (1860–1952), known as "Copey," was beloved by succeeding generations of Harvard undergraduates. "For almost forty years 'Copey' kept open house Wednesday evening after ten in successive chambers. . . . and no Harvard instructor has ever accumulated so devoted a following as his. An excellent lecturer and teacher of English composition, he did not confine his instruction to his classes. The lads flocked to him because he liked them; his conversation ranged over all subjects, and brought out the views of his guests; his readings gave the most illiterate a taste for English classics" (Samuel E. Morison, *Three Centuries of Harvard, 1636–1936* [Cambridge: Harvard University Press, 1936], p. 402).

Nathaniel S. Shaler (1841–1906), a professor of geology, did much to make the Harvard geology department among the top ten in the country. His lectures were "vivacious and speculative. He was frank, hearty, alert; he had a picturesque manner of speaking both in choice of words and intonation—even so simple a monosyllable as 'man' gained a Shaleresque turn on his lips" (*ibid.*, p. 312).

LeBaron Briggs (1855–1934), described as possessing a "happy compound of naturalness and fastidiousness of precision, honor, and grace," was an instructor in English from 1883 until he assumed the deanship of Harvard College in 1891 (*ibid.*, p. 76).

Barrett Wendell (1855–1921) was a professor of English whose personality managed to encompass "a more or less transparent mask of contrariness and eccentricity, united creative power with sound critical sense." That Livermore, a science major, should single out three members of the Department of English is both a commentary on the quality of the department and the importance of the English curriculum to the Harvard undergraduate.

to know anyone "outside," and often carried his club prejudices into post-college life.

My football career terminated ingloriously. I never made the first team after my freshman year, although I was always on the squad. I sat longingly on the bench the one year we beat Yale, a substitute when not one man was retired. My senior year was spent on the second team, then used as "scrubs," who, overworked, tattered and unsung, were yet close to heroes. The last play I ever made in Harvard was a diving tackle which nailed my man, but knocked me out with a concussion which clouded my memory for hours. It is hard to pick out of four years the events that are worth recording. Sad to say, the less reputable ones stand out the farthest in memory, perhaps because the element of daring or risk involved left the most impressions. High spirits of youth made us do a lot of things that oldsters shook their heads at then, the same as now. I remember them with a sort of wicked glee.

Boston, reachable by trolley over Harvard Bridge, or, on more exuberant occasions, by "hack" from Blake's, the long-suffering Cambridge livery, was more of a mecca for being a bit inaccessible, and one was apt to make more of an evening of it in consequence.

There were a good many encounters of the "town and gown" type which seldom resulted in more than a fracas to be talked about, which in these days of gunmen and gangsters wouldn't do at all. There is no atom of sporting spirit in them, and murder would doubtless result. I was good-natured rather than scrappy by nature, but more than once found myself in a row not of my making, as when four of us landed in a restaurant bar in Bowdoin Square for a late snack and one of the party got into an argument with one of the habitués, a tough group of "night hawks" (free-lance cabmen). In a jiffy the air was full of fists, bottles, a knife and a blackjack. I got the blackjack away from the user, and with it as a threat we fought a good rear-guard action to our cab; then in an excess of gallantry I gave back the weapon, like a conqueror generously returning a sword on the scene of battle. We were afterwards told by our driver and friend that we were followed to Cambridge by a cab bursting with toughs, but evidently not quite up to closing for action.

The one secret society I joined (long out of existence) was one which required among other things for its severe initiation (I bear some of its marks today) that the candidate do some deed at which if caught he would be expelled. Men were picked for this society for their supposed adventurous spirit, and I must have qualified on this ground, though I never liked the destructive tendencies which it acquired later and which caused its dissolution. My stunt was to paint the face of the clock in the tower of a prominent church in Cambridge late at night. Paintpot in hand, I found my way through a dark alley to a conveniently open window, threaded my way through aisles of musty-smelling pews, feeling like a sacrilegious wretch all the time; then up the clock tower, and, poking my head out the square window in the face of the clock, daubed its face as far as I could reach, and scuttled for safety. I must have had a conscientious spirit then, even in a poor cause, for, once safe in my rooms, I began to wonder whether I had done a thorough job, and rather than incur the risk of being derelict, did the whole thing over again. This time, no one could have criticized its thoroughness of execution.

Not to dwell too much on escapades, I must complete the sequence by the tale of my final battle of college days, though this was in my last year at Tech, when I prided myself on maturity. Again, however, I took up the cudgels for a friend as much as for the joy of battle. Needless to say, there was a certain amount of "spiritus frumenti" involved. Bill Stickney and I were at the theatre, and a big, youngish, rather flashily dressed man began abusing him for some fancied insult. I stepped in and took up the quarrel, which went on as we walked, as far as the Touraine Hotel, and there I challenged him for a fist duel to take place in the alley opposite. By this time, Stickney had disappeared, and Devens, another college friend, had turned up. He and my opponent's friend were constituted as seconds, and the battle started. I was getting a terrible pounding at first, until some remembrance of science came back, and I did some real execution on my opponent. Still, I was near exhaustion when the brilliant thought occurred to me that he might be, too. "Do you apologize?" said I. "Yes, oh, yes," said he, and I replied, "Well, don't let it happen again," and the fight was mine.

I had a broken nose which carried aluminum splints for a time, and two beautiful black eyes of which traces were present when I walked up to receive my degree at Tech graduation; a good beginning for a serious career!

Though I went the usual round of Boston social events, I was never much of a "fusser," as male butterflies were then called, and still seized every opportunity to get away from formality. Victorian Age girls, probably against their will, were kept on a good deal of a pedestal, and most of them were somewhat of a strain to me. Afterwards, when I was studying in Tech and was glad of the distraction of an occasional dinner or dance, I made some very fine friendships among the debutantes of that flight.

In spite of my failings and failures, I got a lot out of Harvard educationally and socially. I used to say in more sarcastic moods that about all I acquired was a taste for a good cigar and knowledge of how to hold liquor like a gentleman, but of course that applies only to those whom the cap fits. Harvard does put a stamp on a man which may be bad or good or a mixture of both, but it is unmistakable; and, on the whole, I like the stamp.

In the various places I have lived Harvard was often unknown, or if known, her graduates held up to scorn as "stuck-ups," "sissies" or "stay-at-homes." Yale was the mother of heroes. There was some modicum of truth in it in my day. At least a certain type were prevalent enough to give her that black eye. I found myself many a time defending my alma mater, and take little credit in disproving the impression, so far as I was concerned. I think I never came to a full realization of what Harvard means until last year at the tercentenary celebration, when the long and beautifully staged procession of educators in their colorful gowns, from all over the world, descended Widener[3] steps to do honor to her, and the dignity of the occasion brought home to the great audience there her eminence as a standard-bearer of high and free ideals.

After all, why shouldn't I be proud. The name of Livermore, one of my stock, first appears in the catalogue of alumni in the year 1722, and since then there have been many more, perhaps none so distinguished as my father, who, though not an alumnus, was given the honorary degree of Master of Arts in 1910 by President Eliot, with the citation "Thomas L. Livermore, soldier, lawyer, man of affairs, and writer; who almost in boyhood fought in the Civil War; now a profound student of its history; pre-eminent among statisticians of the conflict."[4]

[3] "Joyfully accepted," the Widener Library was presented by the family as a memorial to Harry Elkins Widener (A.B. 1907), whose book-collecting habits even at an early age established him as a foremost collector of his day.

[4] Thomas Livermore's *Numbers and Losses in the Civil War in America, 1861–1865* (Boston: Houghton Mifflin, 1901) is a classic inquiry into the statistical side of the great conflict. He also wrote a highly readable personal memoir on the war entitled *Days and Events, 1860–1866* (Boston: Houghton Mifflin, 1920).

I had hardly received my degree when I came down with an attack of acute appendicitis and spent the summer recovering from what must have been a narrow squeak. Appendectomy was still young in 1900, as I found thirty-three years later, when in a Canadian mining town I was taken with severe pains which the local doctor swore he would have diagnosed as acute appendicitis had not mine been taken out. I recovered enough to be shipped home, only to be operated on a little later. Sure enough, my trouble was found to be caused by the inflamed stump, of which too much had been left, and again, I had a narrow shave. I am one of the few who has had his appendix out twice.

Technology absorbed me for the next three years. I learned the rudiments of mining engineering, and graduated as a Bachelor of Science, again without honors, but with comfortable passing marks. Though I went through calculus in mathematics, that science has always been one of my weak points, and today I leave its handling as much as possible to my superiors in knowledge. My talents, if any, lie in other branches of the profession.

Technology was then housed in the old buildings on Huntington Avenue, and the students lived at home or in lodgings. There was little college life, so called, and little encouragement of such by the faculty. In fact, it was thought that social activities had no place in a technical school. I am glad that Tech now has a home of its own, and a solid loyalty among its undergraduates and alumni. I am sure that father, who was a member of the Corporation for many years, had a great deal to do with that outcome. He was a believer in the need of a great technical and scientific school of national scope, existing for those ends alone, and fought successfully the proposed merger of Tech with Harvard by which it would inevitably have lost its identity. At the same time, he favored the full and independent cooperation between the two institutions which now happily exists. President McLaurin said of him, shortly after his death, in an address to alumni: "Full of years and honor he has gone. His place will be impossible to fill, but the record of his loyalty to Technology will remain as a stimulus to others on whom the burdens of leadership will fall."

Truth to say, I myself had little spirit of loyalty to Tech. I had had my fill of college life, and felt myself beyond its social contacts. I am afraid this was rather typical of the traits that made Harvard men unpopular. All I wanted was to get a mining education and get into the field as quickly as possible, and rather resented what I called the "schoolboy attitude" of students and teachers. Now, of course, Technology is a great school for graduate scholars from all over the world, and that attitude no longer exists.

I remember when the "Miners" were trying to organize a football team and, hearing that I had played somewhere before, asked me to come out. I refused as long as I could, then went out one day and played my old position of halfback. In the first few plays I broke through the inexperienced line and nailed the runner for a loss each time; then, the first time I got the ball, made a touchdown. The team's eyes bulged; but, although they begged me to join, surfeited with football, I never played again. It was a small triumph after my last season of frustration at Harvard.

Technology was not all drudgery; I had some time for my usual recreations of riding and shooting. Bill Stickney, my good friend and classmate at Harvard, who was taking a postgraduate course at Tech, and I used to spend many an afternoon in the uplands and on the marshes of the upper Charles, where we would, if lucky, bring back a duck apiece and have them cooked in the basement of the Victoria, and eat them with a bottle of red wine for appetite sauce.

VII

Europe

In the summer of 1901 I went on a bicycling trip abroad with my brother Harry. We had our steamer transportation and five dollars a day for spending money, so we had to eschew expensive hotels and amusements and make the bikes instead of trains carry us as much as possible. We toured Belgium, Holland, Germany, and Switzerland, dutifully using our bikes most of the time, but at Lucerne got homesick for the sound of English, shipped the wheels and entrained for London. Here, we ensconced ourselves at No. 8 Half Moon Street and, with the exception of short trips to Oxford and Cambridge, were content to spend the rest of our holiday.

Some of the highlights of the trip were: when we took a cab in Brussels and were driven to a very expensive-looking hotel, tried to tell the driver we wanted a cheaper one, and he, nodding understandingly, landed us in front of a garish establishment from whose balconies leaned interested and frowsy-looking females. We fairly pushed him out of the neighborhood. The time in a little town called Rosendahl, when we had been pedalling over miles of tooth-shaking pave', and Harry said: "Bob, I've got enough of this; duty or no duty, let's take the train." In exasperating elder-brother fashion, I refused, on grounds of not giving in to temptation, and words led to a challenge to fight it out. We had our coats off and were about to set to, when we noticed a row of highly interested faces gaping at us over the neighboring wall; the absurdity of two American youths of gentle extraction giving these yokels a free exhibition dawned on us, and we laughed and made up. I forget which way we completed our journey. Then the day when we had bicycled to Cologne, a hot and dusty one as ever was, and, refraining from

77

quenching our thirst until we had changed, sat in a cool cellar in front of the largest, most delicious stein of beer I have ever drunk, downed it almost without drawing breath, and ordered, "Noch einz!" Bonn, where we stayed a week with the Bokers, friends and business connections of father's, who lived in a spacious house on one of Bonn's fine residence streets. The number of meals we had and the amount of delicious food we consumed was a revelation; "Früstück" in the pleasant garden in back, where we plucked plums and pears from the tree for our first course; "Mitmorgenessen," when we were there to have it; "Mitagessen," a gargantuan meal; "Abendessen," another; and to cap the day, a heavy supper before going to bed. I'm not sure but that there were one or two other meals for good measure.

Life in this family was strictly "planmässig." Our playmate was one of the two sons of the family, Alfred, of about our age, pleasant, simple, and a bit stolid. Frau Boker laid out each day for us; so long for a tour of the town, so long to see "der Kronprinz" playing tennis on a specially reserved court on which the public could gaze, awe-stricken, from a distance; and so forth. When the elders took a three-day trip to see their other son matriculate, everything was planned for us for the full time; but, once in the clear, Harry and I took Alfred in hand and led him a gentle but instructive round of the town's lighter pleasures. We were good at the American art of "joshing" and gave Alfred full measure, with no light of understanding in his eye, so that we thought our efforts wasted, until the final day, as we sat in a beer garden, recuperating, he burst out from a long silence: "Bob und Harry,—you know, I think you are fonny fellowss." We felt amply rewarded.

I well remember the long, straight German roads, with grades so gradual they were imperceptible to the eye, yet up with our high-geared American bikes made us work like plough horses, while Germans on their low-geared wheels passed us without effort. The neat rows of evergreen trees in the Black Forest, with never a twig or bit of underbush on the forest floor; the picturesque, close-set villages, tucked away in the valleys, and the first glorious sight of the snow-clad Alps; all familiar enough to everyone who goes abroad.

In England, we loved London; who doesn't? Even in summer when "everyone" was away, it was a place of many delights. I think Harry was not for long lured far away from its Piccadilly, but I went awheel to Oxford, and there met a charming former Oxonian, who typically was spending his holiday in the town, and near his college of Magdalen. He luckily took a liking to me and showed me through the various colleges and closes as only one who knew and loved them could have done. His sport was fishing on the Isis. As I expressed a love of the art, he took me with him and set me up on the bank with a rod, while he moved upstream to another spot. I had no luck except for a few minnows, which I tossed back, and when he returned, and, having asked what luck I had, showed me proudly his catch, a half dozen of the same size I had been throwing away, he was horrified when I told him I had done so. My Englishman, on first acquaintance, thought he had paid me a great compliment, when after I had told him I was a visiting American, said, "I never would have taken you for one." I remember having another setback to complacency when I had stopped for the night in a little thatched inn on the way to Cambridge. A couple of yokels were describing a stranger who had passed through, and, turning to me, one of them said, "He were loike you a bit, but *he* were a gentleman."

Harry and I went back as passengers on a cattle boat, the *Londonian*, returning empty to Boston. In those days they shipped cattle on the hoof, and it was a favorite way of obtaining cheap passage. Our fellow passengers were few, and one of them, a naturalized German American, obnoxious, with his thick accent and continual disparagement of everything American, and praise of things German. We wondered why he didn't stay there, but Americans were used to being high-hatted, and I suppose we considered it a normal attitude.

British Captain Lee was most friendly, and, finding that I had been at sea, often invited me to the bridge, and even gave me a quarter-master's trick at the wheel. I found steam steering gear easy, after the bucketing of a windjammer. We hit heavy weather and thick fog on the Newfoundland banks. Once, when I was at the wheel, a small fishing schooner loomed out of the fog so close I feared we would ram her, but no order came to change course and as we passed her I could look down on her wildly tossing decks and see her crew waving and shouting at us, but whether in anger or not I could not tell.

The first news we had from the pilot was that McKinley had been assassinated by Czolgosz, and our rage burned against all foreigners, including Mr. Katzenburg, our German fellow traveller. We didn't know how "foreign" America was to become in succeeding years.

VIII

Mexico

In the summer of 1902, in line with my career, I went as a non-paid assistant to the San Fernando Mine in Durango, Mexico. Father had joined with Mr. Agassiz and Mr. Shaw of the Calumet and Hecla Copper Company in privately financing the property. To anticipate, it gave them only grief from start to finish and was finally closed down, a total loss. My brother Tom was already there as assistant to the manager. I met him at Guaymas, where we took the steamer *Curaçao* to Altata on the mainland side of the Gulf.[1]

This was my first experience in Mexico. I had taken a course in Spanish in college and was eager to try my skill. I was delighted and somewhat surprised to find at the Customs that my few words were understood. On the steamer, I struck up acquaintance with a returning Mexican gentleman who knew no word of English. We had a delightful conversation on about every subject, in a mixture of Spanish, French and Latin, of which latter we each had at least a smattering. So much for Harvard!

At La Paz,[2] the center of pearling on the Gulf, I was offered by an evil-looking character ashore a huge black pearl, no doubt contraband, for fifty dollars. I knew nothing of pearls and had few dollars to spare, so refused, but often wondered if I missed a bargain.

As we coasted along the bleak, barren peaks of Baja California, I felt the stirring of adventure, and wanted to explore that mysterious hinterland, where it was said water was a jewel and the few inhabitants were hostile. Not until thirty years later did I penetrate it, to

[1] Guaymas is a port on the eastern side of the Gulf of California at the twenty-eighth parallel. Altata is approximately three hundred miles south of Guaymas.

[2] A port on the eastern coast of Baja California, La Paz is directly across the gulf from Altata.

find it intriguing, livable for the well prepared, and far from inhospitable.

Altata, with its thatched huts and palm-fringed, sandy shore, resembled my idea of a South Sea isle. Its main industry seemed to be the shipping of huge sea turtles, which were piled on deck on their backs and gave vent to their misery by mellow bleats. Here, we took a rattly open-car railway forty miles to Culiacán,[3] the capital of Sinaloa, and got together our caravan for the journey to San Fernando. My recollections of Culiacán are the clangor of church bells at all hours, experiments in the Spanish language with sundry *Mexicanos* and *Mexicanas*, and the potency of tequila cocktails, with the ensuing headache.

All travel to San Fernando was by horse or mule. There were no roads. After two days of up hill and down dale, camping where night overtook us, we rode into Huamuchila, the native village near the mill, forded the Yumaia River, and were made welcome by the staff at camp. The establishment consisted of the mine perched on top of a mountain to which a railway climbed steeply the several miles from the mill and smelter in the valley below. All of the equipment, some of it ponderous, had either been packed in on the backs of mules or floated, with many a shipwreck, down the river. The mine had been rich, but the Mexican owners pulled a fast one on the examining engineer, both as to price and ore reserves. A couple of years after my visit it was abandoned. It was, however, a scene of bustling activity during my stay.

I took up quarters with Tom at the mine in a glorified tent perched on a ridge whence we overlooked a sea of jungle-covered peaks, with Fau and Sat, two Chinese servitors, to attend our wants. One of Sat's first duties in the morning was to fill our high boots with hot water, when often a scalded scorpion or centipede would float to the surface! It was the only way to be sure of their absence.

This was the *tierra caliente*, and the semitropical luxuriance of the vegetation was enhanced by the rainy season, which had just begun. It disturbed us little, however, as the mornings were fair and the afternoon rains more refreshing than otherwise, but rust and mold covered weapons and leather.

I had procured a very nice gaited horse through Cheminant, the company's agent in Culiacán, which I rode up and down the mountain between mine and mill, and in my numerous trips around the

[3] Culiacán is forty miles due east of Altata.

country. This horse, however, was plains-bred and on him I had three of the narrowest escapes I ever had while riding; once, when on a steep zigzag of the trail he shied, lost his footing and fell over the edge. I was just able to throw myself clear and catch a bush while the horse rolled over and over down the almost precipitous slope, landing in a thick tree, from which we extricated him unhurt, strange to say, though scratched and snorting. Once, when passing between two low-roofed houses he shied violently under the eaves of one, on which I caught my shoulder and ear, and just missed a broken neck; and again, when on a narrow cliff trail overhanging the rocky river bed he whirled end for end and hung a breathless moment on the edge. After that, I shipped him back to the low country and took to the safe mule.

I was supposed to be at work learning the mining business, and I did work hard enough while I was at it, but, not being under pay, was a good deal of a free lance, and took every opportunity to join expeditions away from the routine of camp life. With the survey crew running lines through the jungle was one distraction, when betwixt intervals of holding level rods for the transit man I found time to watch the varied sights of the *tierra caliente*.

Birds were always my passion, and of these there were hundreds, most of which I could roughly classify. Those that intrigued me most were the green parrots which flew in flocks overhead, looking, with their quick wing strokes, curiously like ducks, except for their blunt heads and short necks. The jacalaca, here called the cuitchi, a jungle fowl of somber color, who perched on the bare limb of a jungle tree, every morning made the welkin ring with his harsh cackle. These fowl are good eating but hard to get; I think I only shot one. They are so closely related to the domestic hen that they are said to be mated with game fowl to fiercen the breed, but I don't know the truth of it.

Our camp was in a little glade at the base of a cliff down which plunged a clear waterfall. In the morning, Robinson and I would edge along the cliff base, get behind the fall, and then dash through it, somewhat battered by the force of thirty feet of water, a giant's shower bath.

One trip was to a place called Cortijos to look at a claim owned by one of the staff and worked by his partner. I don't remember how far it was or how long it took us, but I remember the incidents of the trail, how we stopped at a fine-looking quartz vein outcropping on

SONORA

Hermosillo

Guaymas

CHIHUAHUA

Sierra Madre Mts.

BAJA CALIF.

La Paz

Parral
Santa
Barbara

Cerro
Prieto

SINALOA

San Julian
Guamuchil
Altata Culiacán

DURANGO

the hillside, into which a new tunnel had been driven by a couple of prospectors. They were a bit downcast, as their tunnel had caved into a deeper one, driven by the ancients, and, the ore having all been taken out, their work was but a shell. We passed a travelling troupe of musicians a-horseback with every manner of queer-looking instrument; a band of hunters looking as wild as the game they hunted; and a hog driver whose flock looked as if they would attack us as soon as not.

At our destination we found the partner, one Sam Ray, a man of around seventy, an unreconstructed Rebel who had exiled himself after the war. He had married a young Mexican woman and had raised a numerous family ranging from a babe in arms to twelve, apparently at the rate of one a year. They were all dark as Indians and spoke no English, but were named John, Henry, Betsey, and the like, and understood perfectly the awful mixture of Spanish and English Sam addressed to them. One gem was when the mules were straying a bit, to his oldest:—"John, go busca le mula no let 'em so va," and John, with a "Si, Papa, seguro!" jumped to his task. Sam was immensely proud of his family and of his prowess in producing them. As he proudly remarked, "I don't know when they will stop coming. There's another on the way."

On a rock, basking in the sun, was the biggest lizard I ever saw. I don't know what variety, probably an iguana. He was all of four feet long. At our approach, he retired to a cleft in the rock, from which we tried to pry him with a bar and a hook, but his hide was too tough to be penetrated. A few minutes after we left he was out sunning again. "I'll fix him," said Sam, and got a stick of dynamite and touched it off in the hole. When the smoke cleared away, the lizard was still vigorous and angry, as he well might be, so we left him alone and unbeatable.

At San Fernando I picked up some knowledge of mining and milling, but I must have been somewhat of a problem to Mr. de Kalb the manager, until he finally gave up trying to invent tasks for me, and let me go my own way. My youthful senses were vividly alive to the colorful life of the place, its novelty of scene, and the simplicity and charm of the people, so much so that when I left I had real nostalgia for Mexico as I had seen it.

Even in this remote country there was law, represented by Don Castulo de La Rocha, a true hidalgo, head of the numerous Rocha family, who had almost feudal sway in these parts. Don Castulo was *jefe político*, which means the boss and no mistake. He was an old bandit hunter, one of Diaz' captains of *rurales*, and had cleaned up the country, so that for the whites it was safer than a Sunday school. Once Diaz' iron hand was removed, the picture changed rapidly, but when I was there an American was respected, even if not liked. However, the facile passions of these primitive people made a few troubles among themselves, and an exchange of shots or a knifing or two were not unusual. Once, the company doctor summoned me to help him perform an autopsy on the body of a woman murdered in Embarca village across the river, as no one could be found to help him and the Mexican law required an autopsy, no matter how apparent the cause of death. I reluctantly volunteered. We found the poor thing laid out with candles and a crucifix, and an awed populace surrounding her hut. Her lover, in a fit of jealousy, had stabbed her with a butcher knife through the lung and liver (as we ascertained later), and in spite of it she had lived several hours. I nerved myself for the task while the doctor explored her insides, and managed to survive the ordeal but vowed never again to be a post mortem assistant.

People, and incidents whose sequence I have forgotten, stand out in my memory; Jimmy Rhue, jovial Texan, equally at home in Spanish and English, who taught me native songs, "No me mates" ["Don't Kill Me"], "Asi cual pasan las ilusiones" ["Such Are Illusions"], and joked me about Boston, but was a bit touchy when joked in turn about Texas; Wiesener, he who owned the claim at Cortijos; Saunders, competent mill super; Canfield, who lent me his prize mule later to cross the Sierra; Fischer, transplanted German with all the sentiment of his race, who four years later was hospitable to me in Mexico City; and Captain Eddy, the one-legged Cornish mine captain, who with his timber toe bestrode a mule for every occasion, and rode it underground via the main tunnel as far as he could go; the miners, stripped to cotton pants or breechclout (some of them were Yaquis), chanting to the swing of their hammers. That minor wail and clink of steel deep underground made an unforgettable impression.

There was the strange and cruel custom of tying out a beef for twenty-four hours before slaughter without food or water, "to make him tender"; the man who was bitten by a huge centipede, whose leg, when we—the doctor and I—got to him, was swollen to half the size of his body, with the three imprints of every claw where he had grabbed three times at the creature before stopping its progress up his leg; the coral snake that lay in the path outside our tent, deadly but slow and quickly killed; the tarantula that sped from wall to wall of our tent and exiled us all outside; the Fourth of July celebration, when I tried to shoot off the flowing tail of the camp rooster with my six-shooter, but plugged him too close to the root, so he died, and was mourned by our Chinaman.

Harmony prevailed among the whites, largely, as was the uncomplimentary explanation, because no one had brought their wives. The only rivalries were the healthy ones of work and play, and not those created by social jealousies, as I have since seen so often in longer-settled communities. Lack of ladies did create some temporary alliances in which at parting *la chiquita* was set up for life with a jacal (hut) and a Singer sewing machine, and did not lack for native suitors thereafter. There was a certain fascination in some of the "morenitas," and I confess to having said good-bye with some regret to one Arcelia, who, as I look back on it, probably had me marked down, though unsuccessfully, as prospective donor of her sewing machine.

I had an ambition to return to the States via the land route across the Sierra [Madres] to Parral,[4] both for the adventure, and because I wanted to cross the continent a-horse before the coming of the rail. The Kansas City and Orient was then projected to cross hereabouts to connect with the Gulf. I needn't have hurried. There is no rail yet.

In spite of warnings of bandit-infested roads and other perils, I got together my outfit, a sturdy pack mule, and a fine saddle animal borrowed from Canfield. Don Castulo and his Señora were making the trip to his highland home, Atascaderos, and offered me his company thus far, an offer which I gladly accepted. Brother Tom rode with us up the hot valley as far as the company's ranch, where after lunch we parted. I must say I felt a bit lonely in a strange land, with a strange tongue to wrestle with, and unknown adventures ahead.

Our cavalcade consisted of la Señora, riding ahead, muffled to the eyes in veil and *rebozo* [scarf], and, token of her caste, a superfluous

---

[4] Livermore is referring to Hidalgo del Parral, a city in the southern part of Chihuahua, on the twenty-seventh parallel. From the number of references to Parral in the autobiographies of mining engineers, everyone going to Mexico must have passed through that community!

hat over all; Don Castulo, dark and dignified; and I; then the pack animals in the rear, pushed on by several *mozos* [boys]. We climbed steadily all day, first in the heat and dust of the semitropical valley, then, as we gained altitude, to cooler slopes, until toward evening we reached the summit of the plateau, where great pines grew parklike amid green grass and flowers.

We camped under the shadow of Chonteco's peak, pitching tents for ourselves and letting the *mozos* get what sleep they could by the fire between gusts of rain and chasing uneasy mules. Next morning, on we went, the tropical world we had left at our feet and an upland wilderness of forested crags ahead. The trail followed ridges where possible but now and then dipped into arroyos or zigzagged up their steep sides. One such place I remember, called "la escalera" (staircase) but looking more like a ladder, was as stiff a place for a mule as I have ever seen. The Señora funked a little, but, plucky though pale, kept on and made it without accident, as did we all. Only mountain mules could have done it.

Two days of this, in which I found myself conversing with more ease in Spanish now that I had no English to fall back on, brought us to a level pine forest, open and grassy. Don Castulo turned to me and said with a wave of his hand: "Mis terrenas, Señor" ["My fields"]. Then a jog through the pines, interrupted by a fierce hailstorm which drove us to shelter and whitened the ground, brought us out upon a beautiful prairie, flourishing with cornfields and dotted with ranch buildings. "Es la hacienda de Atascaderos" ["This is the Atascaderos estate"], said Don C. with manifest pride, and the Señora's eyes glistened with the joy of homecoming after a year's absence.

We rode through the arched gate to the patio enclosed by the low walls of the main house, and were welcomed by peons and Rochas alike. I felt a bit out of it while the family greeting and *abrazos* were going on, but with the innate courtesy of these people, they soon made me feel at home, once the formalities of introduction were over. The Don's sister, a stately dame, was really impressive with her "A Rocha at your service, Señor."

Ismael, the oldest son, a few years older than me, took me to share his room, where I slept the sleep of the weary, and was greeted in the morning with "Good mornin' frien'" from Ismael, but this was the extent of his English. His next remark was "Le picaron las pulgas?" (did the fleas bite you?), to which politeness compelled an answer far from the truth.

Nothing would do the hospitable Rochas but for me to stay a day or two, and on the first Ismael and I set out on a deer hunt. We saw only one, which our fusillade from the saddle failed to bring down, but we enjoyed ourselves thoroughly, two young men, riding good horses in a gorgeous country of grass and towering pines, acquainting ourselves by lively discussion of points of view derived from widely different upbringing, yet for all that much alike.

My stay, short as it was, had established me in the affections of these friendly people, and my parting created quite a bustle. Don Castulo insisted on my taking one of his peons, one Feliz Trejo, who proved an invaluable guide, servant and friend. The mules brought into the patio were packed and saddled. The Señora hurried about, bringing from the dairy cheesecakes and tortillitas, tucking them in my saddle bags, murmuring "Don Roberto might like this, Don Roberto might like that." Don Castulo overwhelmed me by placing a beautiful betasseled hair halter on my mule's head "como recuerdo del pais" ["as a remembrance of the country"], Ismael grasped my hand with "Good-bye, friend," his only other English, and his pretty sister shyly seconded him with "Adios, Don Roberto." Don Castulo rode with me a way for manners' sake, and gave Feliz stern instructions to see me safely in, then bade me farewell. I asked him what I could send him from my country, and, pointing to the two-and-a-half dollar slicker oilskin on my saddle, he said, "Ah, much would I like one of those fine mantos, but I fear it is too much to ask." I am glad to say I did not forget either him or Ismael, and hope they received them.

A longish ride through pine-studded fields and sparse settlement brought us toward evening to the adobe walls of Aguas Blancas, the hacienda of Don Miguel Ojeda, to whom I bore a letter from Don Castulo. Don Miguel took his time about reading it, with no word to me; then, satisfied, bade me welcome with "Pase, Señor, la casa es suya" ["Come in, Sir, the house is yours"], and escorted us into the patio, mules and all. Supper was served al fresco, at which I had my first experience in eating a cooked green chili, and though my eyes streamed with tears and sweat poured down my brow, I finished it, to encounter the amused glances of my companions. There were no womenfolk at this place, and after supper we sat at ease in the starlight and talked, or rather the others did, while I listened and trained my ear to the local Spanish, to which I was rapidly becoming accustomed to the extent that I could understand the gist of what was said, though Feliz' tale about *difunto* puzzled me until I was told that it was about a ghost.

In the morning, Don Miguel rode with me a good two hours, perhaps as a special mark of favor to such a good friend of Don Castulo, until, with the remark, "Es el fin de mis terrenas, Señor; que vaya con Dios" ["This is the end of my fields; God be with you"], he turned and left me.

San Julian, a little mining town, was our evening stop. It was *ralla*, or payday, and the scene was a gay and lively one, with crowds of peons in bright-colored serapes gathered around the various buildings gambling, dancing and strumming guitars; and bepistolled horsemen galloping about. Though the staff house where I applied for lodging was manned by English clerks, I got no encouragement, so I sought shelter in the native store, whose proprietor allowed us to unroll our beds on the floor, and fed both us and the mules for a pittance. Incidentally, this was the only time I was allowed to pay money during the whole trip. As we rode out in the morning, an Englishman rode after us and asked me to return, saying they had been upset the night before because of the death of their superintendent, but I was unforgiving and rode on.

Up to now, we had been riding through a country of great ranches, sparsely settled, but easy to travel. From now on, we entered a wilder terrain of forest and mountain cut by frequent swift torrents, through and over which the narrow trail wound its tortuous way. Hardly were we clear of town when a deer crossed our path, at which I shot and missed. Beside the trail Feliz pointed out to me the tree from which recently a notorious bad man had been hung, with a shred of the rope still pendant, and the grave underneath.

Feliz, once sure of his "patron," proved a rare companion. To him, I owed a real proficiency, at least for that trip, in the language of the country and for one song of the mountains, "Quisiera llorar" ["I Would Like to Cry"], which he taught me patiently. He was a true philosopher. I suppose, like all peons, he was a serf to the soil, owning little but the clothes he wore, but happy, tuneful, and contented with his lot. He told me he had been married (but not by a priest, they cost too much) but was no longer. Why? Because, said he, "Guando no hay carino, mejor partir" ["When there is no love, it's better to part"], which, after all, is the general idea today.

Rain descending in sheets caught us in muddy going and dark was upon us before we saw the light of our destination, Rancho Viejo. Here a big fire blazed in the open portal of the log house, and our precautionary hail brought the reply "Pasen Ustedes" ["Come in"]. This was a poor family, but we were made welcome of their tortillas and beans, and to shelter under the portal. There was a sick child, whom I as an *Americano* and, of course, skilled in medicine, was asked to doctor. Taking a chance on it being chills and fever, I prescribed and gave her quinine and a hot toddy of mescal from my pack, and in the morning was told, to my relief, she was better, thus saving the reputation of the American medical profession.

At Cerro Prieto,[5] the village lying under the mountain of the same name, we lunched next day at the dobe castle of the *dueño* of the village, Don Jose Mario Franco, who, hungry for news of the outside world, plied me with questions, thee'd and thou'ed me, gave us a royal meal, and ended by taking such a pointed fancy to my hat, a cheap panama, that I, with true Castilian politeness, gave it to him, and bought myself in the village a steeple-crowned sombrero with which I was more pleased, as being more in character with the country.

Don Jose told us of one "Don Vays" living in camp beyond the mountain, which, if lucky, we might make that night, so on we went through rugged country, passing, I remember, some rock caves in which dwelt naked Indians, who peered at us through tangled locks and scuttled out of sight. Then up and up, until, in the gathering dusk, we rode a narrow trail on top, the world at our feet. I named it, to Feliz' approval, "La Cumbre del Mundo" ["The Top of the World"]. Pitch darkness overtook us in a wilderness of roaring creeks whose boulder-strewn beds we must ford. How Feliz knew where he was going was beyond me, but at long last, in a grassy park surrounded by tall pines, a light appeared and at our hail at the door of a tiny cabin a tall figure appeared who answered "Quien es?"

5 Cerro Prieto is approximately thirty miles southwest from Santa Barbara in the province of Durango.

["Who is it?"] When I spoke, he answered in English "Get down and pass in." "Don Vays" proved to be an Arizonan, Wise Webb by name, foreman of a big cattle outfit ranging these hills.[6] He said he was doubtful of receiving us, first because there were a lot of bad characters hereabouts, and second because his wife was expecting a baby almost any hour. She, a really beautiful Mexican girl, most evidently with her time nearly upon her, nevertheless got us supper of sorts and made room for us to spread our beds on the earthen floor. I must say that I was a trifle relieved when we bade good-bye next morn with no need of my acting as obstetrician. I have often wondered how that young couple fared, alone in the hills, no doctor within leagues.

Not far from here next morning two horsemen driving a band of horses passed us on the lope, pulling up to scrutinize us closely but saying nothing. They were a picturesque couple, in *charro* costume, much silver about their gear, and armed to the teeth. Whether they were cowboys or bandits, I didn't know, and Feliz was reticent, but I was glad of his company, of my own well-armed condition, and the assurance of Diaz' law and order.

Toward afternoon we rode out of the hills to the plains of La Providencia,[7] in the midst of which stood the hacienda of Don Eduardo Franco, a gentleman with a numerous family, whose bare feet were clad in guaraches [sandles], and who, though as usual hospitable, consistently shouted at me to make me understand better, not unlike my own countrymen with those who struggle with English.

Don Eduardo had a grievance. He bade me carry a message to my compatriots that the wolf poison they sold him was "muy malo" ["very bad"]. The wolves took the bait at his very door and thrived on it. When he told me that he used a whole package of strychnine, which of course acted as an emetic, I advised smaller doses and perhaps vindicated my country's reputation. When it came time for bed in the central room where I had unrolled my blankets, I was a bit embarrassed to find that the ladies of the family evidently slept there too and had no intention of leaving while I disrobed, but I solved the problem by turning in with only my boots off.

This was nearly the end of our trip. One more steep and scary

[6] Wise Webb may have been a member of the Webb family in Arizona; see "The Webb Story," *Arizona Cattlelog*, Vol. III (September, 1947).

[7] About fifteen miles southwest of Santa Barbara.

climb brought us to Santa Barbara,[8] wide-awake and wide-open mining town, where gambling booths lined the streets through which we rode without stop; then across the level plain to Parral, on the railroad.

Parral seemed to us, after months in the mountains, a truly metropolitan place, with its well-dressed people, busy shops and streets, the smoke of trains and industry, and, strange to our eyes, wheeled vehicles. I believe I had not seen a carriage or even a cart since leaving Culiacán. In turn, we must have looked a bit out of place as we rode up the main street, clad in hot-country clothes, armed and shaggy. At the "Cosmopolíta," where we applied for lodging, the proprietress fenced a bit, rather suspicious of our respectability, then, looking at me more closely, said, "You speak English, don't you? So do I, I'm American," and thenceforth we concluded our business with ease.

Feliz and I changed to our best, and after supper sat in the Plaza watching the endless procession of young people promenading in a circle, men one way, girls the other, then I went to bed, while Feliz, with his pay in his pocket, saw the town. I bade him good-bye in the morning and watched him ride out, driving the spare mules ahead, his sandal-clad toes thrust in the stirrups, his only purchase, a pair of gaudy shoes much too small for his splay feet, swinging by a cord from his neck, and, I venture to say, as penniless as when he arrived.

Mexico made a very strong impression on me as a land of charm, simplicity, and escape from the conventions of home, that lasted me a long time. Though I have been there often since, I never recaptured the feeling of romance my first trip gave me, nor have I had the opportunity to live so intimately among the people and know them so well. Alas, politics, propaganda and revolution changed them much. A lone American journeying now as I did then would receive far different treatment, I fear.

One memento I had of Mexico was not so pleasant; "calentura," the hot-country malaria, got its grip on me, and at intervals for a couple of years made me alternately shiver and burn. Not until I went to the high country of Colorado did I shake it off.

[8] Ten miles from Livermore's ultimate destination of Parral, Santa Barbara was the center for considerable American investment in Mexican mining.

IX

Camp Bird, Telluride, and the Smuggler-Union

Tech finished, I was offered a job by Mr. John Hays Hammond, with whom father was on friendly terms, as assistant engineer at the Camp Bird Mine at Ouray, Colorado. Mr. Hammond had engineered the sale of the mine from Thomas Walsh to the English company who now owned it, and was acting as their consulting engineer. I can't help thinking how much as a matter of course we took the fact of a job being ready for us then, in contrast to what a young graduate is up against now. If one was willing to leave home, there was work in plenty, and in fact all through my western experience and up to the war, no one needed to be out of work for long.

A year before this my brother-in-law, Bulkeley Wells, had taken charge of the Smuggler-Union Mine in Telluride, and, as my sister was joining him in Denver, we travelled thus far together. A digression here is in order to record the beginning of affairs in which I was to be intimately involved the next few years.

The miners' strike, brought on by the Western Federation of Miners, after much preliminary rumbling broke out in 1901. Every camp in Colorado was to feel its effects, but Cripple Creek and Telluride bore the brunt. The union was ruled by the notorious "inner ring," of whom [Charles] Moyer, [George] Pettibone and [William] Haywood were the heads. Moyer, the president, was supposed to be the willing catspaw. Pettibone was the arsenal, by reason of his bomb manufacture, and Haywood the arch conspirator and

Courtesy of the Western History Research Center, University of Wyoming

Robert Livermore at about twenty years of age.

The Camp Bird mill.

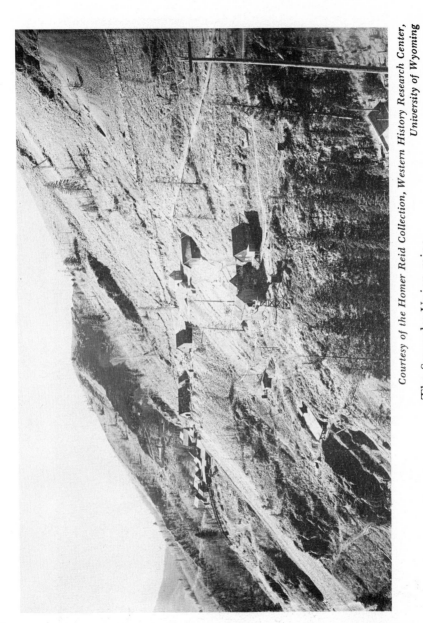

Courtesy of the Homer Reid Collection, Western History Research Center, University of Wyoming

The Smuggler-Union mine.

Courtesy of the Homer Reid Collection, Western History Research Center,
University of Wyoming

The main street of Telluride, Colorado, about 1910.

The Smuggler-Union boarding house.

Courtesy of the Western History Research Center, University of Wyoming

Grace Livermore Wells and Robert Livermore, 1905.

Courtesy of the Western History Research Center,
University of Wyoming

Bulkeley Wells

Courtesy of the Western History Research Center, University of Wyoming

Halstead Lindsley, 1906.

Courtesy of the Western History Research Center, University of Wyoming

General William Jackson Palmer, about 1905.

The Japan and Tomboy mines.

Courtesy of the Homer Reid Collection, Western History Research Center,
University of Wyoming

Street scene in Telluride during the 1903 strike. Notice the militiamen.

The Smuggler-Union mill and cyanide plant in the early 1900's.

Courtesy of the Homer Reid Collection, Western History Research Center,
University of Wyoming

Pandora miners, early 1900's.

Courtesy of the Homer Reid Collection, Western History Research Center, University of Wyoming

Telluride's Cosmopolitan Saloon, about 1912.

Livermore en route from Goldfield to French Mine, Nevada, 1911.

From the left, W. Spencer Hutchinson, Robert Livermore, and Quincy A. Shaw.

leader. Under their command were a band of terrorists, dynamiters, and gunmen, to whom murder was a commonplace. The rank and file of the union was probably held in line partly by the usual methods of intimidation and partly no doubt by genuine loyalty to the cause of labor as they understood it. The grievances were vague and various. Though there was no particular objection to wages and hours, numerous excuses were found for a strike. The real reason was the familiar one of a demand for a closed shop and domination by the union. The slogan of the *Miners' Journal*, the organ of the W.F.M., was "Labor produces all wealth. Wealth belongs to the producer thereof." This all has a familiar ring today.

In Telluride, conditions were particularly bad. Under the leadership of one [Vincent] St. John, president of the local union, the town was completely terrorized and the mines were in a state of siege. Several murders had been committed, and the Smuggler Mine had been assaulted and captured by union forces after a pitched battle in which several men were killed and wounded, the latter including Becker, the superintendent. It was then that Senator Buckley of the state legislature, sent in by the governor to investigate conditions, sent the famous telegram "No need to send troops. The miners are in peaceful possession of the mines."

Arthur Collins was manager of the Smuggler-Union Mine at this time. An Englishman, courageous, obstinate, he refused to be dictated to by the union, and was correspondingly hated by their leaders.

One night as he was seated at a table playing cards in his house at Pandora a charge of buckshot through the window ended his life. Wells, then secretary of the company in Boston, promptly volunteered to take his place. Whatever faults Wells had, lack of physical courage was not one of them.

At first the union, perhaps a bit apprehensive of the effect of their latest and greatest murder, were not actively hostile to the new manager; but when they found that he too refused to truckle to them, the strike went on with renewed virulence. The mine operators, following Collins' death, had imported a number of professional gunmen who were made deputy sheriffs for the protection of the mines and the community. These men, of a type now vanished, were nearly identical with the so-called "bad men" of western legend, except that they were on the side of law and order. At that, I guess some of them had been on both sides of the fence. Some had been stock detectives on the ranges, hunting down cattle rustlers, some mine guards in other strike-torn camps, and one or two at least ex-regular-army men. They were lightning on the draw, looking for trouble by preference, and soon instilled a wholesome respect among the "red necks." Of them, more anon.

When I first went to Camp Bird in June, 1903, the state of affairs in Telluride, just over the range, was a sort of armed truce, with minds at high tension and anything likely to happen. Needless to say, my thoughts dwelt much on my own people (strike or no strike, my sister Grace had joined Wells, first alone and afterwards with all four of her children), and every incident and alarm made me long to be there instead of at peaceful Camp Bird.

To return to my personal narrative, I reached Ouray, hired a saddle horse, rode singing up the mountain road to the main office at the mill, and reported for duty to Mr. Cox, the manager, thence through last winter's towering snowdrifts to the mine in Imogene Basin where I was to live and work for the next year and a half.

The mine buildings, at an altitude of 11,000 feet, were located at the mouth of a 2000-foot-long tunnel, through which we walked to work and, once at the vein, climbed long distances to the various working places. Not until later did the management indulge us in the luxury of man hoists. My duties were those of sampler and assistant to the surveyor, and a thorough breaking-in did my immediate superior, Willard Harris, give me. I got so I could wield a single jack with the best of them, and to this day can moil out a tough sample of quartz better than many a youngster. The work underground was hard, and the office work, maps and calculations, tedious, but I made good at it, and have Harris' patient strictness to thank for the foundation of good engineering habits. I remember my pleasure when after some weeks I received a rare word of commendation from Mr. Scott, the superintendent, and substantial testimony that I was holding down my job, a raise from three dollars to three and a half dollars a day. They didn't pay young engineers very highly in those days.

The scene outside was one of real grandeur and beauty, an amphitheatre of towering cliffs, with their toes covered with slide rock, surrounded slopes thickly carpeted with grass and flowers in the short summer, deep buried in snow eight months of the year. Tall spruce forest still covered the lower slopes, and in the distance, where the valley opened a vista, the vertical-sided top of Potosi[1] towered against the sky line.

I had little time for leisure, but in the long evenings after supper I used to explore the tongues of forest reaching into the basin, flushing an occasional blue grouse or snowshoe hare, or picked a secluded spot for sketching, then still a habit with me. I have a pen-and-ink drawing of Potosi in my journal which took me a month of evenings to finish.

I liked best the outside work, when in summer we toured the surrounding country surveying claim lines or triangulating contours. I earned the name of "the mountain goat" by my agility in scaling cliffs and planting marker flags in lofty spots. Up above us there were little flats, glacial cirques, real mountain meadows, usually with a small lake in the center, simply ablaze with flowers, the beautiful columbine, painter's brush, and many alpine varieties unknown to me. Sometimes in one of these, out of sight of the transit man below, I would lie on the mossy sward, smoke a pipe, hear the bees humming, the sweet call of the white-crowned sparrow, and watch the cottony clouds go by in the sparkling blue air, for the moment my own master.

Some of the work took us to the tops of the ranges, locating the outcrops of the veins or finding boundaries, where we would climb about on peaks and ledges where a misstep would have meant disaster, where snow lay deep on the north slopes, and ptarmigan, unafraid, walked barely aside, like the pigeons on the Common at home. Once, nighfall caught us still on the wrong side of the range, clinging like flies to the steep mountainside. The sun setting behind Potosi lit up the peaks and left the gulches in deepest gloom, so that when we reached the top of the divide we walked along a knifelike ridge in sunlight with black gulfs yawning on either side. A month before I would have been scared by a lively imagination, but by then I was seasoned mountaineer enough to enjoy the experience.

I didn't break into the work without many a blunder. One such taught me something well worth while—to mind my own business. While sampling in the stopes[2] I had seen a lot of moils (pointed hand steel) left around by the timbermen, and at lunch, hearing Scott

[1] A peak in the San Juans.

[2] A stope is an excavation of ore above or below a horizontal level, often in a series of steplike formations. There is some opinion that the word "stope" may be a derivation and corruption of "step."

grumble at the way they were disappearing, I piped up: "I know where there are a lot of them;—in the stopes." Nothing was said at the time, but after lunch Beaton, the foreman, whose one blue eye I had caught regarding me fixedly when I made the remark, hailed me. "Come here, young man." "Yes, Mr. Beaton," said I, "What's the matter?" "Ye know verra well what's the matter. That remark ye made at lunch. Ye dinna want to make any more cracks like that again. I'm responsible for what goes on in this mine wi'out any interference, and I'll thankye to mind your own business henceforth." Fortunately, I saw my error, and apologized in good faith. After that, Beaton and I were good friends and he took a fatherly interest in my progress. Many years later when I was manager of the Kerr Lake Mine in Cobalt[3] I sent for him to become superintendent under me. A fine type of Highland Scot, able, upright and loyal to the end, he died in my employ, and I mourned him and saw him to his grave.

About this time, Mr. Hammond and his partner, A. Chester Beatty,[4] paid us a visit. To us young engineers these paladins of our profession, particularly Hammond, with his romantic African experience as a background, were not of common clay. We were correspondingly delighted with their affability to us small fry. Hammond told us we were to consider ourselves as "of his staff," which is the nearest I ever came to being connected with him in business, though he did me a great service later, as will be told.

I was about to invest some of my modest patrimony in Camp Bird stock, knowing as I did from familiarity with its resources that it was undervalued, and confided my intention to Beatty. He said, "Why bother with an investment which will return interest only; why not buy Esperanza,[5] which will double your money?" This Mexican

[3] Cobalt, a small mining town in the southeastern corner of Ontario on the border between Ontario and Quebec, is not far from the recent Timmins strike.

[4] At this time A. Chester Beatty was a young graduate of the Columbia School of Mines, and a brother-in-law of the famous mining engineer and author Thomas A. Rickard.

[5] The Esperanza mine had a checkered and chaotic history, characteristic of many mining enterprises in Mexico. In 1899, John Hays Hammond had secured an option on the Esperanza for English investors, the Venture Corporation. F. W. Bradley, at that time manager of the Bunker Hill and Sullivan Mining Company in Idaho, made a geologic examination of the Esperanza. On the basis of his report, the purchase price of $6,000,000 was agreed upon. As a check on the Bradley recommendation, another survey was made by Ross E. Browne, whose findings showed that the mine was worth scarcely half of the Bradley

mine was one just taken on by the English company at their recommendation. Much flattered, I did so with half my funds, but with
native caution put half in Camp Bird as I intended. For a year or
so I saw Esperanza stock sink, and Camp Bird rise; but of that,
more anon.

One of the demands of the Union was for an eight-hour day for
mill and smelter men as well as miners. The strike had hung fire a
while pending election, but when the legislature refused to enact the
law, it broke out again and all the miners were called out. In early
September, for some reason connected with secrecy I was sent over
the range to Telluride with a message to one of the mine offices, the
Tomboy, I think.[6] I was told to take a few days off with my "folks,"
so rejoicingly I took a horse, surmounted the 13,000-foot ridge, and
rode down the trail to Pandora, where Wells' headquarters were.

The miners were pouring out of the bunk houses as I passed,
shouldering their bedrolls if afoot, or lashing them to saddles if
lucky enough to catch a stray livery horse, and hotfooting it for town.

At Pandora, the mill was going full time, manned by loyal employees from superintendents to clerks. Wells had determined to mill
out the ore in the bins in hopes of adding a few dollars to the
company's sadly depleted treasury. I volunteered and was put to
work on the stamp batteries in company with a fellow Harvard man,
Halstead Lindsley,[7] who had come out with Wells and was already
broken in as a competent mill man and miner. Halstead and I
renewed an acquaintance which ripened to a lifelong and intimate
one, through which in after years we were partners in enterprise and
adventure at intervals, even to now.

We were practically in a state of siege. The town two miles down
the valley was seething with excited union sympathizers who sent
scouts and pickets out to the dead line drawn by our guards across

---

assessment. Later the Venture Corporation purchased a 51 per cent and the
Guggenheim Exploration Company a 49 per cent interest in a total purchase
price of $3,000,000 (John Hays Hammond, *The Autobiography of John Hays
Hammond*, 2 vols. [New York: Farrar & Rinehart, 1935], pp. 505–515).

[6] The Tomboy mine was located immediately above the Smuggler-Union,
3,000 feet above Telluride. Discovered in 1880, it had its most productive year
in 1894; three years later it was sold to the Rothschild group in London for
$3,000,000.

[7] A lifelong friend of Livermore, Halstead Lindsley achieved success and fame
as a mining engineer.

the road at the property line. Stacks of rifles lined the walls of the mill, and every worker packed a six gun at his belt.

At the end of our shift the more aggressive of us would escort the non-union men to their homes in Pandora, putting out flankers and driving the occasional picket concealed in the roadside ahead of us.

In the front room of Wells' house, grim reminder of union murders, were the buckshot holes in the window screen and the gouges in the furniture where Collins was shot down.

In the morning as we sat at breakfast, word was sent that Bob Meldrum,[8] one of the deputies, wanted to see the manager. "Tell him to come in," said Wells, and Meldrum appeared. A short, stocky man, rather insignificant looking except for piercing blue eyes with that intense expression I have since come to know as those of a killer, talking in the flat tones of a deaf person, which he was, it was hard to believe that he had killed in the line of duty at least a dozen men, and was known and feared all over the West.

One of his first acts on reaching Telluride was to enter the union saloon, crowded with "red necks," march to the bar, shoulder aside two of the worst characters, and announce, "I'm Bob Meldrum. You can always find me when you want me. Now, if any son of a bitch has anything to say, spit it out; otherwise, I'm going to take a drink—

[8] John R. Burroughs, in a sketch of Meldrum for the Denver *Post*, began by flatly stating, "Robert Meldrum was a professional gunman." This killer for hire was born in the state of New York in 1865. In 1899, he appeared in Baggs, Wyoming, where he was employed as a harness maker. After collecting a reward for turning in a "wanted" criminal, Meldrum left for Colorado. Burroughs writes that he was employed by coal mine operators in southern Colorado to keep peace, and later by mine owners in Telluride. He showed up again at Baggs, Wyoming, in about 1908, where he was asked by the Snake River Cattlemen's Association to discourage horse stealing, although some insist to this day that he was employed by sheep men in the same vicinity. Regardless of his employers, Meldrum's tactics were highly successful. By 1910, comparative quiet settled over the Snake River Valley. Two years later, Meldrum indiscreetly murdered a popular cowboy by the name of John "Chick" Bowen. The first trial concluded in a hung jury and he was released on an $18,000 bond; it was four years before Meldrum stood trial again. This time he was given a five-to-seven-year sentence in the state penitentiary. However, his "punishment" consisted of working for ranchers in the Rawlins vicinity. After serving his time, he opened a harness shop in Walcott, Wyoming. Sometime—the precise date is unknown—his shop burned and he mysteriously left southern Wyoming. There are several theories as to his death, but they remain unsubstantiated (John R. Burroughs, "Bob Meldrum, Killer for Hire," Denver *Post*, September 23, 1962).

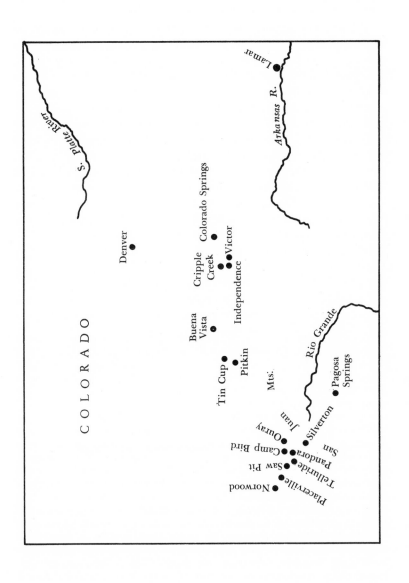

and alone." With that, he contemptuously turned his back, and amid dead silence finished his drink and stalked out through a wide and respectful lane.

Meldrum was, I was told, English born, but little trace of the Cockney remained. Exceptional for a gunman, he was talkative, witty, and, when in the mood, could tell of his killings and adventures more interestingly than a novel. He had a smattering of education, which an occasional half-humorous phrase brought out in contrast to his usual colloquial slang.

The present occasion was a collision with union officials who, claiming that the road through the property was a public one, had tried to drive around the mill. Meldrum stopped them, and, at a remonstrating word from them, jumped into the buggy, dragged one out and slugged the other full in the mouth, knocking out a couple of teeth. They got back to town somehow, and telephoned violent protests to the mill office. Wells cautioned Meldrum not to be so rough. "Why, Mr. Wells," said he plaintively, holding up his hand bleeding at the knuckles, "the bastard bit me!"

In view of Meldrum's past reputation with a gun, and his subsequent performances, the strikers were lucky to get off so lightly, and nothing came of it.

The time came for me to go, and reluctantly I saddled up, rode through new-fallen snow to the top of the range, then turned my nag loose to find his way home, and waded and slid through the drifts to Camp Bird.

The agitators in the Ouray union tried to call a strike in sympathy with Telluride, but the Camp Bird men, who were in the majority, were a canny lot, not easily stampeded, and voted against it. These men were largely "blue noses" from the old mining districts of Nova Scotia, and a good crew. Their home language was Gaelic, and as Beaton was equally versed in that language, they got on famously. One of my recollections is of walking in to work with the men, with Gaelic jests tossed along the line, echoing from the walls of the tunnel like the wild sound of the pibroch. I even picked up a few words, but never could get the hang of that language of imagery.

Life resumed its humdrum way, but my ears were always attuned to news from strike-ridden Telluride, whence rumors of alarm and violence came to us at intervals. Meldrum got his first man in Telluride,—someone who resisted arrest, and whom Meldrum, always cannily inside the law, first incited, then beat to the draw. A new state governor, Peabody, a friend of capital,[9] declared martial law, and the camps filled up with soldiers. The mines, guarded by sentries, resumed work. General Bell, then adjutant general, made Wells captain of a local company made up of loyal men from the mines and mills, of the more high-spirited citizens, and of cowboys from the lower country. These men had high *esprit de corps*, worshiped their leader, and, after the militia withdrew, kept order of such sort that never again was the union strong in Telluride.

On our side of the range were only the ordinary incidents of mining. Two men, drilling a machine, drilled into a "missed hole" and were killed. The mine shut down for the funeral, but I, thinking it a good chance to get samples undisturbed, went underground, the only man in the whole mine. In one of the stopes I noticed a machine standing at a queer angle, and suddenly realized that I was at the spot of the accident. It didn't need the further gruesome evidence of blood and brains splashed on the rocks to send me out of that place in a hurry. A mile underground, alone with such reminders, makes one feel quite isolated.

I was granted a holiday for Christmas, and though winter had set in in earnest, I climbed the range on snowshoes and glissaded down the heights above the Tomboy, where Wells met me with a horse. Thence to Pandora for festivities with the reunited Wells family, where jollity in spite of alarums reigned.

Telluride, filled with uniforms, miners down from the hills with well-filled pockets, cowboys up from the range, their cattle shipped, their horses standing in dozens outside the town's forty-odd saloons, presented a gay and festive sight. It was always a picturesque town, just a bit more gala in the style of the old West than the neighboring towns of the San Juans. Like most prosperous mining camps, it attracted men of many antecedents. Among them were a score of university graduates from Yale and Harvard and a sprinkling of young Englishmen, who lent it tone. The "New Sheridan" hotel was

[9] Livermore's sentiments on the pro-mine-owner biases of Governor Peabody were shared by labor and capital alike. James H. Peabody, born in Vermont in 1852, came to Colorado twenty years later. Interested in public utilities, he was elected president of the First National Bank of Canon City in 1891. Peabody was elected Governor in 1902 and served until 1904. He died at Canon City on November 23, 1917 (James H. Baker and Leroy S. Hafen, *History of Colorado*, 4 vols. [Denver: Linderman Co., 1927], V, 415–416).

famed for its good fare, its vintage wines and imported cigars. On the bar of every saloon was a steaming bowl of Tom and Jerry, free to all; the roulette wheels clicked merrily, and "down the line" Telluride's Yoshiwara was alight with mistletoe and lights, white as well as the accustomed red.

Needless to say, Lindsley and I sampled freely of the Tom and Jerry, lost a few dollars on the wheel, and absorbed enough of Christmas cheer to make a sleigh ride down the valley needful to clear our heads for the home feast, which, with a haunch of bear meat instead of a boar's head, Wells conducted with his usual mastery of quip and song. Sorry indeed was I to climb the lonely cliffs and snowshoe back to Camp Bird.

Martial law kept Telluride pretty much in order. St. John had been indicted and had fled the state. Bad characters were rounded up, and the more notorious of them bounced none too gently out of town.

In February, after a short visit, I was leaving Telluride by train when the state troops were leaving. The populace en masse was seeing them off, cheering everybody, with a rouser for "Captain Wells," when some enthusiast shot off his gun, and the last view I had as the train pulled out was of every horse in sight careering riderless up the street.

The strikers took courage with the departure of the soldiers, and in March returned in force. The newly formed "Citizen's Alliance" took them promptly in hand, however. A scouting party rounded them up along with various undesirables who had hitherto escaped deportation, barracked them in a vacant store over night, then loaded them on a specially ordered train and shipped them out of the country to the accompaniment of the usual warning fusillade. There was a lot of criticism of this no doubt lawless action in outside circles, but Telluride had suffered too sorely to take chances. The vigilantes were made up of the best people in town. Although there was some roughness, there were no shootings, as well might have happened, and it was pointed out that these same men who had killed some and driven others of the loyal employees, ill-clad, over the snowy range afoot in 1902, were royally treated by contrast, with a train for transport.

Violence breeds violence, and word soon came that the Silverton union was arming and preparing to invade Telluride. Martial law was again declared, the Telluride troop was mobilized, and soldiers without number poured in from Denver. Isolated as I was, I didn't hear of all the events in this troubled town, but they were stirring and many. I heard of Wells' troop being ordered to Ophir,[10] a hotbed of unionists, to arrest some ringleaders; and of Wells, impatient at the slow progress of a body of men, spurring ahead, with one sergeant as bodyguard, finding them all at meeting in their hall, appeared at the door, commanded them all to sit still, and held them until the troop arrived and picked out their culprits. Moyer, who had incautiously come in to see what was doing in Telluride, was promptly seized and held incommunicado. His counsel, one Richardson,[11] while pleading for his release, being a bit too emphatic in his remarks about the deputies, was slugged by one of them, and fled the town to ask for a hearing in Ouray.

Lindsley, who had been sent in to guard the attorney acting for Wells and Bell,[12] against whom a habeas corpus for Moyer had been served, wired me that he was to be in Ouray, so I donned skis and travelled to snow line, then rode to town. The place was full of scowling exiles who, seeing my meeting with Lindsley, well known to them, promptly classed me as an enemy. In fact, the local sheriff told me that Richardson had had me pointed out as one of his assailants, and demanded protection from both Lindsley and me. To

[10] The first mining claim in Ophir was staked in 1875. Five years later it was a bustling, thriving mining metropolis with six major lodes. By 1900, Ophir was going the way of many mining towns in the West whose riches, quickly discovered and just as quickly exploited, became exhausted, with the community fading away, sometimes slowly but occasionally almost over night (Muriel Wolle, *Stampede to Timberline* [Boulder: University of Colorado Press, 1949], pp. 391–392).

[11] Edmund F. Richardson, head counsel for the Western Federation of Miners and a veteran of many labor strikes and litigations, served with Clarence Darrow as defense attorney at the Haywood trial in Idaho. See David H. Grover, *Debaters and Dynamiters* (Corvallis: Oregon State University Press, 1964), and John A. Remington, "Violence in Labor Disputes" (Master's thesis, University of Wyoming, 1965).

[12] A colorful character, Sherman Bell had served in Theodore Roosevelt's Rough Riders, was a sponsor of the only Spanish bull fight in the United States, and was an unceasing enemy of socialism. At the Cripple Creek strike of the Western Federation of Miners in 1903, he roared into town yelling at the top of his lungs, "I'm here to do up this damned anarchistic Western Federation" (Marshall Sprague, *Money Mountain* [Boston: Little, Brown and Co., 1953], p. 251).

allay his fears, one of our Camp Bird men got me to meet him, and I did, for the lark. He proved a pompous individual, yet evidently nervous and afraid. He gave me a long dissertation on the wickedness of Telluride, and, in spite of my sponsors' recommendation, regarded me with suspicion, as well he might.

The hearing came off in the morning before a union-inclined judge. He listened with impatience to Wells' attorney, who pleaded military necessity, but with attention to Richardson's bombastic oration, and ordered a writ to be served on Bell and Wells for their arrest, and a fine of five hundred dollars to be paid to Moyer! A naive judgment, to say the least. Needless to say, nothing came of it, except, I believe, that Moyer was released.

Spring days made pleasant after hours in Imogene Basin. We skied on the steep slopes on what I afterwards knew as "corn" snow, punctuated by many a tumble on skis much too long and held on only by a toe strap. Yet, on these same contraptions the Scandinavians working above us on a neighboring claim came down in long and graceful swoops, without a fall.

Once, we climbed to the top of the range, to find ourselves challenged by a lone guard armed with rifle and gun belt, stationed in a hut amid snowdrifts to prevent exiles from returning. With him we fraternized and relieved his loneliness by target practice, then borrowed planks from his wood supply, and coasted home.

In June, Scott called me to his office and told me that he had reliable information that a union plot was afoot to go by different trails to the divide, overpower the guards, and blow up the Tomboy pump house at their supply lake. He asked me if I was willing to cross the range and warn the authorities, as obviously telephoning wouldn't do. I enthusiastically accepted the mission, and, to avoid observation, set out early to climb the thousand-foot cliffs, rather than the more roundabout and visible trail. I made it in an hour, quite a feat as I look back on it, and, challenged by the sentry, told him my mission and was passed on to tell my story to the head office in town. Reinforcements were promptly sent up, and the attempt was not made; whether because of my warning or not, I never found out. I made it back to Camp Bird that night, and slept most of the next day.

On this trip, I learned from Wells that in spite of military protection affairs were in bad shape. Although actual terrorism in Telluride had stopped, constant threats and vile accusations kept everybody in a state of nervous tension; yet there was no tangible enemy to fight. The strike had spread to other camps, and miners were leaving the state by hundreds. In these circumstances, the unlucky company was losing money rapidly, and it was only a matter of time when they would have to shut down; in fact, the shutdown came late that month, somewhat to the relief of all.

Meanwhile, Cripple Creek had been getting the brunt of it. Following many beatings and shootings, the superintendent and foreman of the Vindicator Mine were blown to pieces by a bomb set underground,[13] and worse was yet to come. Hell was popping in that camp, where for the time being the union had the upper hand, and murder stalked unafraid.

About this time, one of the vicissitudes of mining happened to me; a piece of steel flew from the drill I was using and lodged in the exact center of my eye. Fortunately, it didn't pierce the cornea, or I would have been minus one eye, but it gave me trouble enough. Continuing to fester and to blind me, it at last forced me to go to Denver, where a remaining fragment was extracted by magnet.

Grace, hearing of my trouble, rose to the emergency, and, meeting me at the junction, gave me welcome company; and in spite of a damaged optic, I thoroughly enjoyed with her, after the strain of high altitudes and strike warfare, the pleasant society of Denver and Colorado Springs for a few days. Among other incidents, I remember my first ride in an automobile, when we toured the Garden of the Gods, actually without a breakdown, a memorable experience.

On return to the mine, Wright, the bookkeeper, proposed to Harris and me that we should go shares in grubstaking a prospector whom he knew and thought well of, one Mart Flanagan, to find us a lease among the mines of Cripple Creek. We took him on, and

[13] The responsibility for the Vindicator explosion, on November 21, 1903, has never been determined. Charles H. McCormick, superintendent, and Melvin Beck, shift boss, were killed instantly as they descended into the shaft of the mine. For varying viewpoints of responsibility, the following may be perused: B. M. Rastall, *The Labor History of the Cripple Creek District* (Madison: University of Wisconsin, 1908); Western Federation of Miners, "Reply of the Western Federation of Miners to the 'Red Book' of the Mine Operators Association," *Miners Magazine*, 1904; Colorado Mine Owners Association, *Criminal Record of the Western Federation of Miners—Coeur d'Alene and Cripple Creek—1894–1904* (Colorado Springs: Colorado Mine Operators Association, 1904); Irving I. Howbert, *Memories of a Lifetime in Pike's Peak* (New York: G. P. Putnam's, 1925).

shortly he informed us that the Independence Mine, one of the Camp's formerly richest, was shortly to go on a leasing basis, and that if any of us could pull the strings we might get a favorable block of ground. Promptly I thought of Hammond, who was consultant for the property, and wrote him for an introduction. Shortly came a letter from the management saying that a block had been allotted to me, and in great glee I got leave, and set out for Cripple Creek.

We met Flanagan, a fine type of western miner and prospector, and together we went down the Independence shaft, to find that we had been allotted probably the worst and most unpromising block in the mine, far beyond our means to develop; so we sadly returned to Ouray, whence I sent a letter of thanks and regrets to Mr. Hammond, and forgot the matter in the routine of work resumed.

Cripple Creek had just emerged from the worst of the strike and was like a battleground torn by shot and shell. Following the Independence Station horror,[14]—where fourteen workmen waiting for their train were blown to bits by a bomb set under the platform, whose gory evidence I saw on shattered, blood-stained walls,—a meeting of citizens in the square at Victor was addressed by Clarence Hamlin, leader of the mine operators. Shots rang out from the union hall, but the bullet intended for him killed a man at his side. Soldiers were at hand and besieged the union men in their hall. They felt secure behind their brick walls, but high-powered Krags punctured them like paper, and soon drove them out to be made prisoners. Meanwhile, the enraged populace had made wrecks of many houses and stores known to contain union sympathizers.

Not having been a participant, my chronology of events may be incorrect, but my remembrance is that the strikers were loaded on freight cars and shipped under guard to the Kansas line, then told

[14] Most contemporary evidence indicates that the Independence Depot explosion, unlike the Vindicator mine bombing, was the work of the Western Federation of Miners. When Harry Orchard confessed to the crime, there seemed to be do doubt. However, recent evaluation of Orchard's testimony would indicate that there should have been considerable skepticism concerning Orchard's veracity, as his testimony is shot through with prevarications and distortions (Remington, "Violence in Labor Disputes"). Clarence Hamlin, a former partner of Verner Z. Reed, who sold the Independence mine at Cripple Creek to British investors, was one of the most able of mine operators. Brilliant and hard-driving, Hamlin had more respect in the Cripple Creek district from both employers and employees than has frequently been admitted.

to hoof it over the prairies and never come back. The few days that I was there were lively ones. "Red necks" were still being rooted out, and twice, under the sponsorship of my friend Will Blackmer, with whom I had been in South Pass in '98, I was invited to join a posse to deport some malcontents. I was all willingness, but in each case, to my disappointment, word came to disperse without action.

Life seemed tame in Camp Bird after the excitement of Cripple Creek, but in August my skies were brightened by receipt of a letter from Mr. Hammond, quoting a telegram he had sent to the management of the Independence, saying: "Do all you consistently can to aid Mr. Livermore in securing the desired lease," and urging me to go over there and run it if successful in getting it. Shortly, this was followed by advice from Victor saying that I had been allotted a block on the third level, which was in the thick of the productive area. So much for a helping hand. A story hangs by this letter from Hammond. I kept it by chance, and years later found it and put it in the family scrapbook. In 1935, Mr. Hammond, then resident in Gloucester, called on us at the farm in Boxford, which led to a pleasant renewal of acquaintance for the year or two before his regretted death. As a matter of interest, I showed him the letter which marked one of the turning points in my career, and expressed some doubt, since it was initialled, whether it was his signature. "By Gad, I'll make it so," said he, and forthwith put "O.K." against it, with the date, and signed it with his characteristic flowing hand.

This time I bade farewell to Camp Bird for keeps, and with good-byes and good lucks all said, set off on horseback for Ouray and the train. But on the road I met Hal Lindsley and a group of picturesquely clad friends who urged me to take in Telluride for a day or two before parting; so, nothing loath, I turned about and rode back with them over the range.

It had been decided to lease the Smuggler Mine, following Cripple Creek's example, as being a solution of the strike problem, for men will work for themselves or a small employer under conditions against which they would rebel on company account. Lindsley had secured the principal lease, and with me at work and money in prospect, Telluride was lively again.

This seemed to be the occasion for a parting celebration, so a group of us—a couple of Yale men, three of us from Harvard, and an Englishman or two—decidedly hit the high spots of Telluride's bright lights, and when we galloped home to Pandora in the early hours we had left a trail of rejoicing in about every roulette and faro game in town, not to mention the champagne parlor of the famous "Pick and Gad."

Next day, when tapering off at the scene of our late festivities, tragedy lent its local color. One Umstead, who had tended bar for us the night before, had had bad blood with one Leaming, a cattleman and so-called "tinhorn" gambler, primed himself with whiskey, armed himself with two frontier-model colts, and announced he was going to "get" Leaming. He found him in the Tremont saloon, but made the mistake of talking first, and Leaming, watching his chance, shot him in the back, then pumped several more bullets into him as he lay on the floor. The ominous sound of shots and of smoke issuing from the swinging door of the saloon signified to Halstead and me, standing outside, that someone had "got his," and we breasted the stream of panicky out-comers, to find Umstead gasping his last, and Leaming rolling a cigarette, saying "No one can threaten me with a gun and get away with it."

With this depressing but characteristically Telluridian episode as a parting memory, I bade good-bye and once more crossed the range, this time by way of the rugged Virginius Pass. I led my weary nag over ice and snow to the valley below, rested the night in peaceful Ouray, and left for my new activities in Cripple Creek.

X

Cripple Creek and Goldfield

Arrived at Cripple Creek, I threw my energies into making the lease
a go. At first, it looked pretty hungry, but Flanagan was a tower of
strength, and gradually ore began to open up. I, being the only
capitalist, even though a small one, had put up five thousand
dollars to carry us through to production. It was touch and go.
Many a night Flanagan and I ourselves shovelled a carload of ore
from the bin to catch a monthly settlement. After a month or two,
we began to make money, and for the year we had the lease never
failed to make a monthly profit. I believe we realized a profit of
something like fifty thousand dollars all told, but as it was divided
among four, it was no bonanza. A shrewder capitalist would have
exacted a larger share for the use of his money, I guess, but I was
never much at this kind of bargaining. At any rate, I had plenty of
income from it, according to my then simple tastes.

In spite of the recent purge, Cripple was full of agitators and
malcontents, some who had escaped deportation, and some who had
leaked back. The headquarters of the mine operators and better
element was the Cripple Creek Club, which I was soon invited to
join. Within the club was an organization known as the "Rough
Riders," all deputy sheriffs, ready for action at a moment's notice,
and of this also I soon became a member. One day we were told to
report armed at the club. I found everyone there bristling with
weapons and "raring to go," but alas, we were told we were to act but
as a reserve, and that as a matter of policy the miners from the mines
were to take on the actual job of cleaning up.

Someone said, "Here they come!" and down the street came
columns of workmen, silent, grave of face, dinner buckets in hand,

108

belt guns on hip, an impressive sight. They marched quietly to the union headquarters, overpowered the brief resistance made by the deputies placed there on guard as a legal show, then rounded up all inside and held them while the town was ransacked for other known bad ones. These were added to the throng and all driven out of the camp and told never to come back. The Rough Riders, myself included, were stationed at strategic points to guard against rescue. A regrettable incident that night was the looting and destruction of the union store, for which there was no excuse. No matter how great the provocation that brings on vigilante activities, it always turns loose the worst elements to discredit whatever good has been done.

The leasing system generated a crowd of free and easy mining operators, always ready to take a gamble, and with enough real money made to make success seem always around the corner. We worked hard and played hard. The camp was wide open, though at intervals spasms of virtue shut down gambling—never any other vice. Though the principal mines were near Victor, five miles distant, that town never had the prestige that Cripple had, whose National Hotel received countrywide notables, whose restaurant and saloons supplied ample and varied refreshment; and for those whose tastes were riotously inclined, the "row," where the hetaira of the camp held sway, furnished untrammelled liberty of action.

I had a little buckskin pony which I rode to the mine each day, and more than once on Saturday rode the forty miles to Colorado Springs for a change from mining circles. Like as not, with the help of a borrowed mount, I would play polo on him Sunday and ride back the next day. He was a tough little animal.

Once or twice a year a crowd of us would go to the great reservoirs near Lamar, Colorado, for duck shooting.[1] In spite of the comment in my diary of that date, which I now read with amusement, "fair shooting for these days of preserves and decreasing numbers of ducks," we probably had such shooting as will never be seen again. For an average five-day shoot, we sent back 258 redheads, and all of these were distributed and thoroughly appreciated among our friends in camp. Besides these, we killed many more which we used on the spot. To those who enjoy duck shooting, a description of that paradise of the sport may be apt. Ne-No-Sha, the largest lake, six

[1] Lamar is in the southeastern corner of Colorado, thirty-five miles from the Kansas border.

miles long, is where all the ducks lie by day in calm weather. In the evening, or if it becomes rough, they fly to the smaller lakes over known courses which after a trial or two give excellent shooting of the pass variety. No decoys are needed. In a really high wind, all that is necessary is to lie in a hollow in the prairie, when flock after flock swings over, fighting the wind and barely clearing the grass. This was remote prairie country then, a few scattered ranch houses, the caretaker's house at which we stayed, the rest uncultivated plains. I am afraid conditions are far different now. They were good fellows, those leasers and Rough Riders, whether at work or at play, Jack Price, Doc Cunningham, Billy Dingman, Frank Clark, and others; many a tuneful and memorable trip we had together.

Continued peace in Cripple hinged upon the right sheriff, Bell, being elected. The subversive element was still strong in numbers but lacking in office, and they were making every effort to elect their men. On election day the Rough Riders were all on duty at the polls, and although nothing happened in my precinct, a man was killed and a couple wounded elsewhere. Toward evening, as returns began to come in, the crowd gathered at the club, and as it became evident that Bell was winning, enthusiasm could not be restrained. A band was conjured up, a procession formed with the Rough Riders at the head, stores were scoured for brooms, and, joined by the whole well-affected populace, we paraded first through the better part of town, then through the hostile quarter. War whoops were not enough here; someone let off a gun, and soon a fusillade which sounded like a full-fledged battle followed. Windows banged shut, and lights went out. That part of town was for the time a city besieged.

Colorado Springs being so near, we had a good many visitors from that city of leisure. It happened that two of my eastern friends, Bill Rogers and Phil Livermore,[2] bringing with them a delightful French-man, one Comte de la Borde, arrived election evening, and, much to their pleasure, were included in the Rough Rider end of the procession. When the firing began, I was amused to see de la Borde, at first startled, enter fully into the spirit of the occasion, pull out an unsuspected miniature revolver and contribute his share with a "pop, pop, hoorah! hoorah!"

[2] Philip Livermore was a "very distant cousin" who lived in New York City (Mrs. Gwendolen Livermore to Gene M. Gressley, October 30, 1964).

I took quite a shine to de la Borde. He had a real sense of humor, and was altogether likable. I took him underground at the Independence, and introduced him to my partner, formally: "Mr. Flanagan, let me introduce Count de la Borde." "What is the name?" said Mart. "Count de la Borde." "Please to meet you, Mr. Count," said Mart. "A great pleasure to me, Mr. Flanagan," replied the Count with composure.

Shortly after this, as things were going well at the lease and I was rather hungry for a glimpse of home, I decided to go east with my three visitors. Somewhere on the plains, the train was stopped by some small mishap and I walked forward to investigate, leaving the others in our drawing room leisurely waiting progress. In a spirit of fun, on return I burst into the room yelling "Sauve qui peut!" The expression on de la Borde's face and his instinctive leap toward the door makes me smile still. It speaks well for his sense of humor that he burst out laughing before he had taken two steps. A month or so afterwards he asked me to join him in a lion hunt in Africa, but that confounded N[ew] E[ngland] conscience prevented my doing so.

Although Boston had its charms, and displayed its usual friendly hospitality to one of its approved sons, I stayed only long enough to make the rounds of my friends, try out some of my old fields of sport, and spend Christmas at home. Once more, I felt the tang of New England air as I downed a partridge over my fine dog Dazzle, now getting old. Again, I shot black ducks with Eli at the "point of the medder, jest dark" (our total score was four; shades of Ne-No-Sha), and loved it all, but the pull of a lustier world was strong, and joining Lindsley, who also had come home for Christmas, we left for the mines.

We stopped off at Calumet and Hecla in Michigan, and were shown all over that great property, then queen of the copper mines. The "porphyries"[3] had not yet been born, and African copper was not even dreamed of.

After a few pleasant days with the Wells family in Colorado Springs, I donned winter riding clothes, saddled my buckskin and made for Cripple Creek through a foot of new-fallen snow. The lease, under Mart's efficient management, had done well, making

---

[3] Colloquially, any igneous rock that outcrops in sheets.

$3800 for the month, with plenty of ore ahead. My prompt reaction was that the world was my oyster; I took leases on two other properties, put men at work, and spent my time supervising them as well as at times putting in my round of shots along with the others. At first, prospects looked good, but soon one of those runs of bad luck that seem to have no end considerably dampened my optimism. One lease was flooded out; the other, whose vein at first dazzled us with a thousand-dollar assay, pinched to nothing. A man was badly injured by a premature blast, and, to cap the fun, on an 18-below-zero day, while walking home, I was greeted by a miner with "Your ears are froze, pardner." Sure enough, they were, and absolutely without feeling. Rough treatment by rubbing with snow at last restored the circulation, though for a time I thought I would have to go earless, and that night I felt as if I wore a pair of elephant's ears. I had a bad time of it, with raw and infected surfaces, and finally gave up and went to the Springs for recuperation.

I seem to have been rather at loose ends following this episode, and no doubt, nothing loath, was persuaded that my convalescence should be topped off by a spring shoot (then legal) at our old stamping grounds of Ne-No-Sha. We had our usual splendid shooting and flooded our friends in Cripple and Colorado Springs with birds. Among them was the lady who afterwards did me the honor of marrying me, and she told me that if she had received just one more unplucked duck, the cook would have quit. I don't know how many other households were similarly affected.

On this occasion I learned what I had suspected, how poor a shot I was. My blind was on a point, with the wind blowing directly upon it with increasing strength. The redheads came by in tens and twenties and always sagged within good range as they came opposite. I held on the leader with what I thought was plenty of lead, and invariably killed the bird three or four behind him. I still have plenty of that fault.

I went back to the Creek fully recovered, and worked hard for a while underground at the Burns lease on Bull Hill, but, try as I would, no ore materialized. Fortunately, the Independence kept up its steady earnings and, ably managed by Flanagan, required little attention from me.

Halstead Lindsley, who incidentally was doing very well on Smuggler, and I had gone into partnership in a small way, he with me on the Burns, and I with him in grubstaking a prospector and leaser in the new camp of Goldfield, Nevada.[4] One day he telephoned me, "Bob, let's get out of this miserable weather, take a trip to Los Angeles where we can bask in the sun, and take in our lease at Goldfield on the way back." That met my idea exactly, and off we went as soon as we could meet in Denver.

On the train we met a persuasive individual who painted in glowing colors a mine of which he had the handling, in Sonora, near Hermosillo.[5] The mine sounded attractive and the climate warm, so we were easily persuaded to extend our trip thus far south, and, after a short side trip to look at an older silver mine in Lake Valley, New Mexico, joined our friend in Hermosillo.

We found the country warm in more than one sense, as one of the more violent Yaqui[6] rebellions was going on, and no one was allowed to go on a trip without an escort of soldiers. After a look at the soldiers, we decided we wouldn't be any too safe with them, and as we could get no promise of a definite time, we gave up the idea of mine hunting.

We saw no actual fighting, as it was mostly of the ambuscade variety at some distance from town, but the evidence of a severe putting down of a rebellion was at hand in the usual postal photographs of Indians hanging from trees, in the bands of prisoners kept at work building new prison walls for themselves, and, more pathetic, the crowds of women and children penned up in the public square, whose men had either been killed or shipped to exile in Yucatán.

An American couple whom we met had adopted a little Yaqui girl of three or so, whose parents had been killed. I remember she had the most liquid eyes; in them, the wild, startled, yet confiding look I have seen in those of a tame doe. Utterly Indian, she had already forgotten all but a word or two of Yaqui and was beginning to talk and act like an American child. We speculated on what would become of her, growing up amid an alien race, with no connection with her own.

[4] A delightful and sparkling reminiscence of Goldfield society at the height of the mining boom is in Sewell Thomas, *Silhouettes of Charles S. Thomas* (Caldwell, Idaho: Caxton Printers, 1959).

[5] Hermosillo is roughly 450 miles south of Tucson.

[6] The Yaqui unrest was part of the political upheaval and turmoil which marked Mexico's history for the first two decades of this century. The Yaqui revolt is outlined in John W. F. Dulles, *Yesterday in Mexico* (Austin: University of Texas, 1961).

Coincidence has set my surmises at rest, for three years ago, while on a mining trip with an engineer from Los Angeles, I told the story. He said, "Were the adopted parents so and so," naming the people we had met. He then said "Well, you don't need to worry about her. They moved to L.A., where she grew up, studied music, and is now happily married."

Los Angeles and southern California in general were then widely advertised as the American Riviera, but we did not find the "dolce far niente" [the good life] we were looking for. Los Angeles was then a small, bustling, typically American city, rapidly spreading out over the surrounding farms. We found no palms or sand there, but, after a thirty-mile trolley ride through fields interspersed with real estate developments, at last reached the beach. After one look at the cold, grey Pacific, we decided to do our basking and bathing elsewhere.

Somewhere along our travels, we had received a telegram from Blake, our Goldfield partner, "Struck it rich—come at once." But far from hurrying us, this message confirmed our intention of doing up the coast thoroughly, with the pleasant anticipation of a golden reward at the end.

We stopped in Santa Barbara, then a sleepy, picturesque town just beginning its career of wealth and fashion, drove in a carriage over the dusty, rail-fence-lined roads of Montecito, where a few easterners had established quiet country seats, and as in duty bound took a dip in the still grey and cold Pacific.

San Francisco was our next stop. That city will always have an attraction, but the old city, before the "fire," had an indefinable charm. It was a blend of the old and the new, with its Barbary Coast still active; its old Chinatown, quite a different place than today's neat quarter; its famous restaurants, Tait's, the Poodle Dog; its fine hotels, the Palace representing the best of the old, the St. Francis, brand new and up-to-date; its mansions on Nob Hill, built from the wealth of '49 and the Comstock; its harbor full of shipping, with more square-riggers to delight my sailor's heart than I had ever seen before together. There was an atmosphere of devil-may-care, of free and easy ways and of gaiety that made it a fascinating stopping place for a traveller.

We took in the races, the Britt Nelsen fight, lunched on the piazza of the old Cliff House, and watched the hundreds of seals on the rocks opposite, sampled a bit of San Francisco night life, enjoyed sumptuous meals at the St. Francis, then entrained for Nevada.

A crowded narrow gauge train took us from Reno to Tonopah, jumping-off place for the new gold fields. Tonopah, itself a prosperous silver camp, had all the paraphernalia of a gold rush.[7] Streets ablaze with lights, tons of freight piled up along the sidewalks, picturesque crowds filling every saloon and gambling hall, excitement in the air. Morning saw us ensconced in one of the few motor stages, which in the thirty-mile ride across the desert passed every form of transport known to the West, from the prospector with his burro, pack trains, twelve-mule freight wagons, six-horse stages, many of them of the old Concord type, to the 1905 edition of automobile in which we rode; a contrast of the old West with the new which will never be seen again.

We found Goldfield as yet largely a city of tents, but with buildings springing up, and seven thousand people already the nucleus of its rapidly growing population. Among the crowds on the streets, we saw many of our whilom antagonists, the deportees of Telluride and Cripple Creek, which reminded us that here the W[estern] F[edera-tion of] M[iners] had found refuge and was at its old methods of terrorism.

We hunted up Blake for a look at our bonanza. "Boys," said he, "I'm sorry to tell ye that the streak petered out right after I sent that telegram." "Never mind," said we, "we had a good time spending it." Truth to tell, though, our hopes were rudely dashed, and perhaps we looked with a jaundiced eye at other prospects in the camp in consequence.

The Combination mine was then the bonanza,[8] and it, with two or three others, were rich, and no mistake. It was said that men cared

[7] Located by James Butler in 1900, Tonopah reached its peak in 1902, when the Tonopah Mining Company was organized with eastern capital. Soon 20,000 miners were scrambling over the hills in the vicinity. A dozen years later Tonopah was well on the way to becoming a ghost town (Richard Lillard, *Desert Challenge* [New York: Alfred A. Knopf, 1942]).

[8] In the spring of 1903 one Al Meyer and R. C. Hart took over a few abandoned claims, which they referred to as the Combination. On May 24, 1903, they struck their first major lode and by September they were shipping their first ore; from then on for several years the Combination was one of the largest producers in the Goldfield district (Muriel S. Wolle, *The Bonanza Trail* [Bloomington: Indiana University Press, 1955], pp. 350–351).

little for wages, they made so much more in the high grade they carried away in their pockets. Later, they even had specially made pockets to carry more, but even the union thought it a little raw when they began loading their dinner buckets. It was a high-grader's paradise, and little was done to stop it, mainly because the leases were short, and everyone wanted to get all the ore possible above ground without interruption by strike or controversy.[9]

Outside the original discoveries, it looked to us as if prospects of making more finds in that monotonous desert were not bright, and we were not tempted to linger. Our steady paying leases in Colorado seemed far more secure. Thus lightly we passed up a chance to be in at the beginnings of what was probably the last rich gold strike in the United States.

Another reason we didn't care to stay too long was because here the shoe was on the other foot. The union was very much in power and had vowed vengeance on their oppressors. There had been several beatings and driving out of camp of those they had on their black list. Lindsley was known to be on the list, and I may have been. But too much money was floating around for them to bother with us at once, and we made our rounds of the properties without molestation. Just the same, our six guns tucked neatly in specially made trousers pockets felt very comforting.

At night the main street was such as one sees only in the movies now. Hastily thrown up frame shacks mixed with tents, housed banks, saloons, and merchandise. A felt-hatted throng jammed the brightly lighted places, drinking, gambling, and talking of the latest finds. Stacks of twenty-dollar gold pieces, with silver dollars for lesser stakes, served as counters on the roulette and faro tables.

That was my only glimpse of Goldfield in its heyday. If our lease had turned out well; if we had stayed a little longer; if, if; who knows! At any rate, we turned our backs on it and returned to our respective jobs in Colorado. When I next saw Goldfield in 1910 it was a dying camp; it had a short life and a merry one.

On return to Cripple Creek, I found that my outside leases had developed no ore, so I shut them down. The Independence lease had made its usually steady production, but apparently, covetous eyes had been cast upon it, for when I applied for the customary renewal, I

[9] The stealing of high grade ore was as characteristic of mining camps as rustling cattle was of the open range.

was told to see the management. Hammond was no longer connected with the company, and authority had been vested in the English superintendent. He is dead now, so let him rest nameless. The conversation went something like this: "Mr. Livermore, you have done good work on your lease, and we would like to renew it, but you must remember you have had handsome profits out of it, and other people would like a turn at it. If we renew, will you be willing to pay something additional?" "You mean," said I, "higher royalties to the company? They already run up to sixty per cent, and the company is making more out of it than I." "No, the royalties are high enough, but I mean something in the nature of a bonus." "Oh, you mean some cash to you," said I, as it dawned on me. "Well, before I'll pay a bribe to hold a lease that our own efforts have made worth while, you can take it and be damned. Some money costs too much to make." And on that note I left the office. In due course, we were notified that our lease had lapsed, and Mart and I sadly paid off our men, sold our equipment, and departed. Looking back on it, I think I was too noble.

I didn't care much for a while. The bank account was comfortably full; I had a good horse, a fine bay named Monte, and I was always happy on a horse. I explored the surrounding countryside alone or in company, always with an eye out for other promising leases. Often I would ride to Colorado Springs, where, among my other friends, I had a pleasant acquaintance with a sport-loving Englishman named Bevan, with whom and his equally equestrian wife and daughter the plains were scoured for coyotes and jack rabbits, and many an exciting run we had behind his fleet and efficient pack of greyhounds.

XI

Colorado, Nevada—Frustration

In July, I was offered a job of taking charge of a drilling operation on the San Juan River, near Bluff, Utah. R. E. Cranston, head man of New England Exploration Company's placers in California, had optioned a large tract for them, and proposed to drill in hopes of finding new dredging ground.[1]

I knew nothing of placer work but opined I could learn, and joined Cranston and his crew at Mancos, Colorado, to which point a Keystone drill had been shipped and was being assembled for the hundred-mile trip from rail.

Nowadays, good motor roads make short work of that trip, but it was a different matter then. We had an eight-ton cumbersome machine, geared for travel with a boiler-run engine, capable of perhaps six miles an hour and a couple of wagons with fuel, water, and supplies. For greater mobility, scouting out camping places, errands and the like, Cranston and I rode horseback, and led a pack pony with our bedding and light camp stuff.

The journey to Bluff took us nine days in all, and every mile of it was earned. The route lay down McElmo Creek for forty miles or so, then crossed twenty miles of desert to the valley of the San Juan. So long as we were in touch with the ranches scattered along the McElmo, we could call for assistance in case of breakdowns, and they were many, but once we struck across the desert we were on our own. The last twenty miles was a forced march.

[1] The vicissitudes of this experience are detailed in Robert Livermore, "Prospecting for Gold on the San Juan River," *Mining and Scientific Press*, CIII (August 5, 1911), 161–162.

The sandy roads soon wore down the teeth of the pinion connecting traction and engine, and our motive power failed. We had to fall back on actual horse power by improvising a wagon tongue, and hitching on hired teams more and more as the road got poorer. When we left the jumping-off place to cross the desert, a spring called Yellow Jacket Wells, we had ten head of horses, and found them none too many. The chief difficulty was that the more horses we had, the more horses it took to haul feed and water for them. One more camp, and we would have been hauling feed for the haulers. As it was, I think we had something like seventeen animals attached to our caravan.

Five miles out, the wagon tongue broke, and we spent the day fixing it, sending all the stock back to the wells. Cranston and I camped in the desert by the machine while a wandering thunderstorm made lightning pictures in the distant clouds. I had shot a rabbit and a few doves, which we had for a game supper. While eating it by the firelight, a Navajo, with his wife, rode silently in, and greeted me in Spanish, full of curiosity at this strange sight on the desert. We offered them the remains of our supper in the frying pan, which they started to eat with gusto. "What is this meat?" "Pajaro" (bird), said I, whereat they spat out their mouthfuls and stopped eating, then, after a word or two together, silently mounted their horses and rode away. Later I learned that the Navajos would eat nothing with feathers, one of their many taboos.

We were told that there was a spring of good water some twelve miles farther, described as being up a draw a few miles off the road near "ruins." We found it, following a steep little canyon which at its head formed a kind of amphitheatre, all fresh and blooming, a vivid spot of green amid the dust and alkali of the desert. In a great cave formed by the wearing away of the sandstone strata were the ruins of cliff dwellings, and in the back of the cave, which was so large that we rode our horses within, was a spring of clear, cold water, whose trickle made the greenery possible. Above, overlooking the wide plain, was a great masonry fort with a round watchtower, all built of carefully squared stone, so well preserved that the ends of the cedar joists which had supported the floors were still imbedded in the walls.

We lay and drank, rested and drank again, grateful for this gift of the ancients, and could imagine this little spot teeming with village life before whatever calamity it was overtook them.

A long pull brought us to the final hill, to ascend which we hitched on our entire force of fourteen horses, and as we surmounted it just at sunset, we could see the muddy river rolling between lanes of bright green cottonwoods, a sight for tired eyes. At the river we found camp, and with one accord stripped for a swim in the rushing stream, a change at least of alkali dust for river mud.

The next day's march, fifteen miles down the river amid shady cottonwoods, no settlement except an occasional Navajo hogan, was a joy ride in contrast to what we had passed, and, as we neared the town, took on something of the nature of a circus procession, with an entourage of men, boys, and Indians, afoot and ahorseback, and all the populace turned out to see us enter. Wide-hatted Mormon girls passed fruit out to us as we rode by, and bade us help ourselves to more. Many of these people had never even seen a railroad, and our machine was a nine-days wonder.

Bluff was then a Mormon settlement of some twenty years' standing, built, as is that people's habit, for permanence, of stone for the most part, with a school and church, all surrounded by luxuriant orchards and fields. There were perhaps thirty families resident. Across the river lies the Navajo reservation, from which there was a constant passing to and fro, as Bluff was a trading post and meeting place for the tribes. Tall red sandstone bluffs rose above the town and the river, and in the caverns eroded from their faces were the occasional ruins of the cliff dwellers.

While on the subject of cliff dwellers, I was immensely interested in their relics. As yet, so far as I am aware, the ruins around here had been little explored, certainly not exploited. The Mormons were uninterested, and the Navajos, superstitious as always, left them strictly alone. Of course, the more accessible ruins had been combed over, and little was to be found except shards, but I imagine there were still interesting finds to be made in the dwellings perched on inaccessible shelves. There was one such of which I made a sketch, a long-walled building in good repair, protected by an overhanging roof of sandstone, a sheer drop below it, for which I could see no possible access from below except by elaborate preparation.

On one tall cliff near the road, six miles down river, was a real museum piece of "rock writing." I was so interested in this that I sketched a page full of the pictographs, goats, deer, squirrels, figures of men, symbols, most of them showing the artist touch. One figure

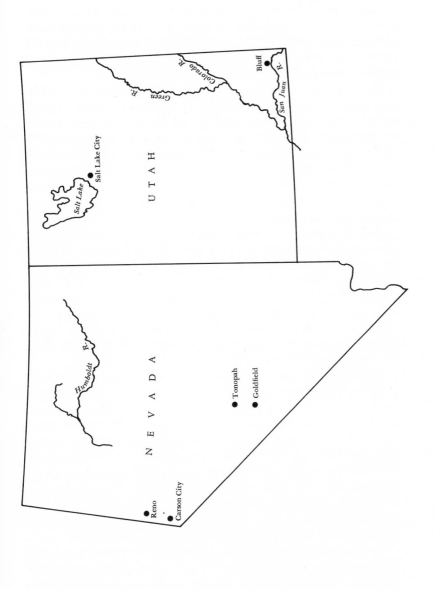

in particular occurred frequently, a man with horns, and otherwise well furnished, playing a pipe, reminiscent of the Greek god Pan.

The Navajos, whom we came to know passing well, seemed to me an admirable tribe. Here as yet there was little of the railroad town influence. They were generally handsome, with rather thin, straight-featured faces, not the broad-faced, high-cheek-boned type of the Utes and plains Indians. They were independent, good-natured and, when occasion arose, hard workers. I had several of them on my payroll, and found them steady, and worth more than the silver dollar a day I paid them. Among them, I remember one "Smiler," so called, I suppose, because he was so preternaturally solemn. He was something of a beau, and at intervals would announce "no work, two, three days," paint his face bright red, don all his finery and betake himself to the dances which went on at intervals.

When in full costume, the Navajos were quite resplendent, clad in a loose blouse, cotton pants slit to the knee, soft tanned buckskin high moccasins, and a gaily colored band around their clubbed hair. Their full complement of jewelry, which they often wore, comprised a silver bead necklace with the conventionalized squash blossom and pendant crescent, a belt of great silver discs which hung loosely on their hips, rings and bracelets set with turquoise and garnets, and often a silver wrist protector against the bow string which they still used.

Their language is rather nasal and full of obscure sounds, yet musical. As nearly as I could pick them up, "ha go sha" meant "where are you going?"—the equivalent of a greeting, "toh," water; "shih," me or I; "shlihn," horse; "ninny," man; "azzy," short; "nochi," Mexican; "nesn," want; and "chiniaga," food. Before I left, with the help of a little Spanish I could make myself understood fairly well.

An arrival of a group from the back country always interested me. The squaws, clad like their men in cotton, much jewelry in evidence, leggings below their skirts, carrying their papooses tightly strapped in cradles on their backs, would gather around in a circle and laugh, gossip, and compare babies much as their white sisters might. The men would drift over to the post to start the long and leisurely job of trading, their stock in hand the fine blankets woven by their squaws and the silver work of their smiths. Days would be given over to this, during which a canopy of boughs would be erected which served as a central lounge. A horse race was sure to be staged, and gambling among themselves or with the ever attendant Piutes, who are the Yankees among Indians, would usually mulct the more enthusiastic Navajo of much of his earnings.

They would take nothing but silver in pay, and this they promptly converted into jewelry. They could never understand that a dollar would buy twice the amount in silver, but no doubt they are more sophisticated now.

While waiting for the freighters to bring in the wagons which we had temporarily abandoned on the desert to use the teams in the last haul, we lodged with one Deacon Allan, one of the local pillars of the church. He was an elderly Scot, well read, a bit of an ornithologist, in which we found common ground. He took quite an interest in me, I think with an eye to gaining a convert, and said I reminded him of Bishop Smith. He lent me a book called the *Story of the Book of Mormon*, a strange hodgepodge of Biblical-sounding names transported to the shores of America. The old sinner had at this time two wives, one his "auld wife," who lived across the street, and Sarah, the active one, with whom we boarded.

Sarah's cooking was beyond compare, her table always loaded with good things to eat. At our first meal we were about to fall to with gusto, when from the Deacon came the words: "Good God!" Startled, we looked around, our forks suspended in air, to find the heads of the family bowed in prayer. Then, after a long pause, came the rest of the grace in broad Scots accents. "We thank thee for the f'ud we are aboot to receive," etc., for a good five minutes, and this began and ended every meal.

One of Sarah's duties was to prepare and take to the other house meals for the "auld wife," and this she hated to do. We could hear her at times grumble and threaten not to, but each time the counter threat of "If ye dinna, I wi' na ca' ye on the other side" brought her to terms. This was a potent weapon, for each Mormon wife, so we were told, is given a heavenly name at marriage by which, and only which, she may be called to heaven by her spouse when they pass to the hereafter. I suppose if the poor thing dies first, she has to wait in outer darkness till her lord and master arrives.

Usually at the open door were two or three Indians, waiting patiently till the meal was over, and invariably they would be given a heaping platter of food. The Mormons are always kind to the

Indians. They believe they are the lost tribes of Israel, whose skins have darkened because of their sins, and that eventually they will be redeemed.

We drilled one hole for Cranston's benefit, then he and I rode the hundred miles back to Mancos, which, unburdened by equipage, we made in two days and a half. I went as guide, and for last-minute instructions, also for odds and ends forgotten at the start. Then back, in company with the claim owner on whose lands we were drilling, to find our outfit camped well clear of town. I thought nothing of a two-hundred-mile ride in those days.

From then on, we lived in tents, taking turns at cooking, and busy from dawn to dark drilling the sands of the river for the elusive values in gold. I panned the products brought up by the sand pump, and weighed the minute specks of gold, but found little to encourage me.

Having no motive power, we depended on Mormon teams to move us from one location to another, and these moves were apt to be the most arduous part of our job. Bog-downs were frequent, and in such case there was nothing to do but dig under and jack up the ponderous wheels, then build a road under them to firmer footing.

Finding little values on our side of the stream, I determined to move across a wide expanse of waterlogged sand to the center of the channel. Everyone in Bluff said it couldn't be done, that the machine would sink into the quicksand, but I persisted. We built a road of sorts across the pools between us and the center sand bars, then rode our horses over the soggy quicksand until the water was squeezed out and more or less dry sand appeared. All of Bluff and the Indians in the neighborhood had turned out to see the fun, and, seeing what we were doing, a few rode down to help us, then more, until a stream of Mormons and Navajos were riding back and forth over our road, to such good effect that we soon had a hard-packed boulevard. With ten horses coupled on, we rolled out over the highway thus made, without a stop, and soon were installed in the comparative safety of the high sand bars of the river's center.

Busy as we were, we had some time for amusement. For sport and to augment our larder, I made myself camp hunter, and drew a steady toll from the teal and other ducks which were beginning to congregate in the back waters. Once, we rode to town for a dance. Dancing being a tenet of the Mormon religion, this one opened with prayer; and for a touch of local color, a row of stoical Ute Indians in tribal costume lined the wall. For all that, the caller's "gent around the lady, and do si do" was as hearty, and the healthy-looking youngsters had as good a time, as at any country dance among the gentiles, and the closing prayer cast no damper on the home-going spirits.

Early in September, I made up my mind that there was no use in drilling more, and left on horse for Mancos to telegraph my conclusions. I slept in the haystack of a missionary who kept a sort of school for Indians, had breakfast with seven little Navajos who chanted grace in unison, and made Mancos the next day. Receiving word to pull up and come out, I started at noon, rode late into the night, and, driven to cover by a thunderstorm, slept in soggy blankets under another haystack, then forded the McElmo in flood and jogged all day across the desert. I shan't forget the pleasure of that ride, alone in a world of rolling brown plains and twisted buttes, red in the sunset light and deepest black in the shadow. The moon gave me light to the river, and ten miles amid the cottonwoods, with the flare of distant Indian corn-dance fires for company, brought me to camp and a long-desired bed.

All was over but the journey to rail. I hired all the teams in sight, and the Mormon boys, commanded by one of their elders, made fair progress to the only watering place in the desert, where we camped and turned out our twenty-six head of horses to graze.

In the morning my two saddle horses were gone, so, telling the Deacon to keep moving, that I would soon catch up, I set out afoot, bridle in hand, to track them down, confident of finding them soon. But Indian pony herds had blurred the tracks, and by the time I had circled and cut them again, I realized I was afoot in a thirsty desert. I simply had to find them, so went at it Indian fashion, puzzling out the occasional sign of shod among unshod hooves, and at noon, six hours later, finally came upon them standing as exhausted and discouraged as I, on the steep banks of the McElmo. The stream's alkali-laden water only aggravated our thirst, and that bareback ride fifteen miles to the Yellow Jacket Wells was a nightmare. There was no sign of the outfit nor their tracks, so after a rest I rode back along the road and found them only a few miles advanced from the morning camp. Hungry, tired, and boiling mad, I let my temper get the best of me, and laced into the Deacon. His explanation was that they had feared I would get lost, and waited for me; and I made sultry remarks to the effect that "no two-for-a-cent Mormon desert could lose me," and that "when I told them to go ahead I meant it," all of which didn't add to the good feeling. To do the Mormons justice, they really were at a loss when after an hour or two I didn't turn up, but instead of sending one of their good trackers after me, they simply did nothing. My tirade stirred their pride and resentment to good effect, however, for without a word they turned to and hauled far into the night.

Arrived at Mancos, I stored the drill, paid off the still surly teamsters, packed my pony with my camp stuff and a choice collection of Navajo blankets and silver gathered during my stay, and set off across country for Telluride. Although I would soon be jobless, I cared little, for the bank account was perforce untouched, I had news that the Esperanza stock, which I had given up as a bad investment, had suddenly risen to twice what I had paid for it, and above all, the skies were clear, the country ahead of me open and boundless, and I had a good horse under me. What more could a man want.

The road follows a great bend of the Dolores River, but I struck across country and rode all day through a virgin land of towering pines and grassy glades untouched by man, to camp on the Dolores far above my starting point. Ice on the pools and a tarpaulin stiffened with frost in the morning reminded me of the approach of fall in higher latitudes. Lindsley had driven out from Telluride to meet me, so the last miles were accomplished on wheels, a luxury after a summer in the saddle.

XII

Wyoming Sabbatical

Arrived at Colorado Springs, I found a letter from "Pat" Sherwin, my boyhood friend, then living at the Forbes ranch in Beckton, Wyoming,[1] renewing plans for a hunting trip in the Rockies. There was nothing in sight in a mining way, and Esperanza continued to rise. Though urged to sell, I said "No, let it ride. I'm going hunting; when I get back, if it's up I'll sell and pay for the trip; if it's down, I will at least have had the pleasure of feeling rich while I'm away." To anticipate, I was out of touch forty-five days, and on return found a stack of telegrams from my broker urging me to sell all the way up to six pounds, which it eventually touched. I sold the stock, for which I had paid a pound and had given up for a loss, at over five pounds, and netted a profit of something like $12,000 for my investment of $2,500. I was playing in better luck than I realized in those days.

My first experience in Sheridan, where I joined Sherwin, was a rough one. I tried out a fine, sturdy young horse, green but apparently gentle. The seller said "he might buck a little," so I touched him up with my spurs to see how little, and found his crowhops in Sheridan's main street easy to ride, and bought him. On the way to the ranch, a loose plank over a culvert made him plunge, the pack pony I was leading pulled back, and without meaning to, I dug my spurs in. With that, my steed exploded. He "buried his head," and punctuated

[1] West of Sheridan, Wyoming, the Forbes ranch was established in the early 1900's by W. Cameron Forbes, onetime Governor of Pennsylvania, and by members of the amazing John Murray Forbes family, entreprenuers in the China trade and American railroading. The Forbes ranch is still being operated by Mrs. Waldo Forbes, whose Red Angus breed of cattle are famous the nation over.

126

each jump with a bawl, gave me the worst ride I ever tried to sit through. I would have weathered the gale, but a ditch under foot caused a sudden swap-ends, and I was lying in the dust wondering how it all happened. Though every joint ached, I had to try it again, and mounted with some trepidation, but fortunately for me, his steam was spent, and I stayed in the saddle. Except for that one blowup, "Reddy" proved a good horse, and served me well.

We got our five pack and saddle horses together and headed into the Big Horn Mountains for a hunt, but found that cattle and sheep had grazed the forest bare, and a new fall of snow made camping difficult, so we packed out of the mountains, bound west for the main range of the Rockies, and for three days jogged across the Bighorn Basin. This was then a dry and barren plain with at rare intervals a ranch and a small town or two. We saw a few antelope but paid them scant attention, bent as we were on reaching more worthy hunting grounds. We reached the little town of Burlington[2] in the Greybull valley,—where Mormons were settling and beginning to make the desert bloom in their usual industrious fashion, long before the discovery of oil filled the land with gentle derricks,[3]—and camped in the dark by the roadside, to find ourselves by daylight nearly in the middle of the town's main street, a bit too public even for that remote place.

We pulled into Cody at last, two hundred miles from our start, and, having decided that we would save time in finding game by engaging a guide, persuaded one Fred Chase to go with us. Fred was somewhat dubious of going alone, as he was used to parties who had a guide for each man and a camp cook, but, looking us over, allowed that we were not regular "dudes," and thought he might take the chance of our doing our share of the work. Before deciding, he suggested a preliminary hunt for antelope, in company with two

[2] Burlington, a supply depot for early settlers in the region, was named after the Chicago, Burlington and Quincy Railroad and located in the Greybull Valley in northwestern Wyoming.

[3] A reference to the famous Oregon Basin oil field, just west of Burlington and south of Cody, Wyoming. The first wells in the Oregon Basin field were drilled by the Enalpac Oil and Gas Company (a subsidiary of the Marathon Oil Company) in 1911 and 1912, but the field did not come under full development until the 1920's. (D. R. Hewett and C. T. Lupton, *Anticlines in the Southwestern part of the Big Horn Basin, Wyoming* [Washington: Government Printing Office, 1917], pp. 181–188).

brothers named Workman, on the plains, a day's ride north of Cody. We set off down the river, then called Stinking Water, a literal translation of the Indian name from the Sulphur springs, more graphic than its modern name of Shoshone.

We camped in the open and the first morning found our tarpaulins loaded with six inches of new-fallen snow, but managed to get breakfast and to cinch our stiffened saddles on touchy horses, and spent the next three days riding after antelope, so few in number, and so wary that our long-range shots failed to bring them down. I imagine that this trip was in the nature of a tryout of our camp manners, and apparently they were found satisfactory, for on return to Cody, Fred announced that he would be glad to go with us.

On October eleventh, having gotten together our pack and saddle outfit, we pulled out in company with a party of four local sportsmen including the two Workmans, who were headed for the same country, and moved up river for two days to the end of road, thence, in the midst of a snowstorm, up a steep and slippery trail toward the Ishawoa Pass, the gateway to the elk country.[4]

We camped perforce to rest our weary stock for a day, though continuing snowfall made the outlook for crossing the divide about timberline doubtful. Here, in need of meat, we circled and cut the tracks of a small band of deer from which I cut down a small and tender buck, which warmed our ribs at least, though the poor horses had to paw the snow to reach the scanty grass.

If we hoped to make it over the pass we had to push on, and, in a blinding snow, came out of timber, to find deep drifts through which we had to plunge and take turns at breaking trail. One of the other party, a rather weak sister, after some hours of this lay down in the snow and announced he could go no farther, but I took him in hand, slapped him to warmth, put him on my good horse, and bestrode his sorry steed until we reached the other side and easier going.

We got to the shelter of timber, and made camp on Pass Creek, though still in several inches of snow,—by contrast a haven of refuge. Here again, a fat deer fell victim to Fred's and my somewhat wild shooting, and fed our always hungry party. Though a few elk tracks were seen, they were old and pointing downhill, so we packed up and followed the ever thicker tracks of the migrating animals to the junction of Thoroughfare Creek, and made camp in a gracious pine

4 Ishawoa Pass is three miles southwest of Cody on a secondary road, Wyoming 1501.

grove overlooking its broad, willow-grown, beaver-dammed bottoms. In front of us rose steep pine-clad mountains topped by the peak called Hawk's Rest; behind us, the rocky ranges of Yellowstone Park, the forbidden ground.

We separated for the first day's hunt, Fred and I crossing the creek on the ice, formed where a beaver dam had made still water. As we crossed, three coyotes, of the timber variety, as big as wolves, slid out of the timber, and though I shouldn't have risked scaring real game, I couldn't resist a shot, and dropped one in his tracks, a beautiful piece of fur.

On top of the ridge we found fresh tracks of a lone bull, who fooled us by doubling back before he rested, spotted us far, and made off. Then in Hawk's Rest Pass we came on the tracks of a small band, and after short and silent trailing, saw them in timber 150 yards away. I fired at a big bull, hit him hard, then ran in, only to see him rise and start off at a good clip. I bowled him over with another shot, but he wouldn't stay down, and I followed, breathless, running and shooting, until a lucky shot brought him crashing down for good, —my first elk. He was a six-pointer, but unfortunately had lost a horn in a fight or an accident, and the head was worthless for mounting. Nevertheless, we "scalped" him, and with backloads of meat, hide and horn, slid and tumbled down the mountainside to camp by firelight. The others had had varying luck; Sherwin none, the others had blundered into the middle of a big band and had killed three, needlessly, as their heads were poor and most of the meat was wasted; but at any rate, we didn't lack for elk meat thereafter. The meat of a young bull or a cow elk, being grass-fed, is much like beef, and one doesn't tire of it as is the case with venison.

We were near the eastern line of the Park here, and discovered it, a straight path hewn through the forest, plain notice to all that this was Uncle Sam's ground; hunters keep out! I remember Bridger's Lake[5] just outside it, a crystal-clear sheet of forest-surrounded water, teeming with ducks, and in their midst three beautiful swans which took wing with loud trumpeting at sight of us, their plumage snowy white against the dark green of the forest.

Pat and I lay near the line one evening, hoping for overflow elk to come our way, but except for a tall moose which stalked by us with a grunt, secure in his faith in the game laws, we saw no quarry. Fred

[5] Bridger's Lake and Thoroughfare Creek are just outside the southeast boundary of Yellowstone National Park in Teton National Forest.

and I, weary of gameless days, repeated this a few nights later, and just at dusk a cow and calf, elk, followed by the bull, walked into the open and grazed toward the line. It was a race between the dark and the shot, but luck was against the bull, for just when another minute would have made our sights invisible, he stepped across and fell, a few yards from safety, to Fred's rifle.

On the twenty-second, realizing that the main herds had gone to better grazing, Sherwin, Fred and I, joined by Charley Workman, basely deserting his party for our livelier company, packed up and hit the trail for the Buffalo Fork of the Snake River. Our road lay over Two Ocean Pass,[6] where at the summit water flows toward both Atlantic and Pacific Oceans. On our side, there was much snow and hard travel, but once over the divide, we dropped into a hunter's paradise. Tall grass bare of snow covered the slopes; springs of clear water gushed out under the pines, and game soon made its appearance. First, a bull elk trotted slowly ahead of us, stopping often to look, as if to tantalize; then three or four big ones crashed down the hillside, with tossing antlers. Still farther we rode almost into the midst of a band, at sight of which Pat and I loosed our ropes, and for the sheer joy of it chased them over fallen logs and sunken streams, but failed to get near enough for a cast. Then as we neared Buffalo, a huge band dotted the slopes of Limestone Mountain like grazing sheep. Why didn't we shoot? Because, for one thing, that whole district had been recently made a state game reserve, and for another, we knew the elk were on the move, and would soon drift near our camp in shooting territory, and save us the trouble of packing in their meat.

At sunset we forded the clear stream and made permanent camp on a miniature sagebrush-covered prairie, bare of snow, and lush with grass. For the next week we hunted in a game country such as I shall probably never see again.

Our first experiences were disappointing. I saw a lone bull standing in a pine thicket and killed him with one shot, only to find that his horns were misshapen, and that he had been crippled by a recent shot, so that his meat was worthless. We scalped and hung his head up, nevertheless, annoyed to find that hunters had been ahead of us

6 At Two Ocean Pass in Teton National Forest, Two Ocean Creek divides, with one fork emptying into the Snake River and eventually the Pacific, and the other fork into the Atlantic.

in the apparently empty land. The next day we came on the tracks of a big band but behind them tracks of horses, and shortly a hunter with pack horse carrying a big head. He said his party was still following the band. Presently, we saw the carcass of a magnificent bull lying in the open, untouched, except that his "tusks" were gone. "Damn it," said we, "The tusk hunters are in the country, so good-bye elk."[7] Tusk hunting was then on the wane. The Ancient and Benevolent Protective Order of Elks had formerly adopted an elk tooth as their insignia, and the result was relentless hunting of the best and biggest bulls just for the sake of these ugly trophies, which retailed at about ten dollars apiece. In fact, I have seen them passed for currency for this amount at bars and gambling houses. To do the Order of Elks justice, when they realized the consequences they changed their insignia, and the worst of the slaughter stopped, but so strong is habit, elk teeth were still being worn and to some extent prized.

We went to camp very much discouraged, but we need not have been, for the tusk hunters went elsewhere, and fresh bands of elk came in and filled the morning air with their bugling. This was an ideal elk country and may be yet for all I know, though I hear of plentiful travel through it nowadays. It was a land of rolling hills, some grassy and open, some thickly timbered, affording plenty of cover and feed, untouched by sheep or cattle. In the distance to the west, the Tetons raise their jagged, snow-covered peaks to the sky, and at their foot stretch the distant rolling plains of Jackson's Hole. A beautiful wilderness, far from the world of railroads and towns, we had even then the feeling that it was all too soon to be parked, regulated, and made accessible to everyone.

We hunted a day or two, separating and riding through the open spaces, on the lookout for a big head. Cows, calves, and small bulls were plenty, but the patriarchs were cautious and kept back in the

---

[7] Tusk hunting continued as late as the second decade of the twentieth century. Jackson Hole was an ideal refuge for the tusk hunter in the winter. Sparse human population, plus a heavy elk population, made Jackson Hole the hunter's paradise. Weakened by a lack of food, the elk were easy prey; a quick slaughter followed by a rapid extraction of the tusks with a pair of pliers, and the job was done. Detection was next to impossible, and prosecution was extremely difficult. The tusks were then marketed in Spokane, Salt Lake City, Denver, or even as far away as New York City (F. M. Fryxell Collection, Western History Research Center, University of Wyoming).

timber while their harems tried the opens for danger. I finally downed a nice six-pointer, whose head I now have, by a cautious stalk through the long grass, dressed him and hung his hide and head on a tree, then put for camp at dusk, chorused in by coyotes. Chase also killed a bull with bladelike horns.

I shall never forget the day of days, October 26th it was. A light snow had fallen; Chase and I had ridden out to get our spoils, which lay not far apart. When I came to the crippled bull I had killed three days before, I saw that a bear, and a huge one by his tracks, had been eating the carcass and had not been long gone.

"Shall I track him myself?" thought I, and was eager to get forth; then caution prevailed, for I knew little of the habits of grizzlies, and didn't know where the chase might carry me. Also, the thought of company when and if he was brought to bay was comforting. Perhaps also I didn't want my partner, Chase, to miss the fun. So I rode back to where he was skinning his elk, and told him the news. Fred's eyes bulged when he saw the tracks. "Let's go git him!" said he. "Why not?" said I.

We planned our campaign. The trail led uphill through spindly "pole patches" of pine, a bad approach to a bear, but there was little choice. Fred thought that he would not have gone far but would be in his bed digesting his feast. As there was no wind to betray us, we decided to follow the trail directly, to get as near as possible by quiet stalking, and to track him on sight, whether there were trees to climb or not. I checked the cartridges in my rifle, shoved my gloves in my pocket, and misquoted, "Lead on, McDuff!" as Fred said in an interview when he got home, "Just like he didn't know my name!"

We had gone about a quarter mile and were in a dense thicket of poles about twenty feet apart from each other, when I saw Fred raise his gun and fire. Immediately, there was an angry roaring which had an almost human sound, and a thrashing of brush, and I saw a big black bulk apparently making toward us. "He's coming!" yelled Fred. Then I, who had held fire to see him more clearly, ran in and blazed away, and so we poured the lead into him whenever we saw brown through the sights, the bear all the time roaring, falling down and rising again, sweeping aside the slender pines as if they were matches, until with our guns empty, his roars subsided to moans and he lay still.

We reloaded and moved forward ready for action, but bruin's days were done. He lay there in the snow, a huge bulk of fur, blood, and froth, a grizzly, and one of the largest. He had created terrific havoc in the brush and had moved toward us about ten feet, but whether he saw us and was charging, so thick were the trees that I can't say. At any rate, he didn't run away, nor did we. We found nine or ten bullet holes in him; indeed, except for the trees, which deflected some, every shot would have told at the sixty feet between us.

We shook hands and yelled with joy, danced a war dance around the biggest game in America, then rolled up our sleeves and went to skinning our prize. Four hours' hard work completed the job, and lashing hide and head on a pole, we packed them out to the horses, transferred them from our aching shoulders to a pack saddle, and made for camp, where we arrived too tired to rejoice much or to admire the six-pointer Sherwin had killed almost from the tent door.

This bear was identified as "Old Four Toes" (he had lost one toe in a trap), a brown grizzly who had roamed this country long and who had a record of having killed two men, though whether he did, no one knows for certain, as they were alone when they met their fate. He measured eight and a half feet from nose to tail, and so broad was he, but little less from claw to claw. He may have weighed eight hundred pounds or more. His hide now reposes in the "gun room" at Boxfield.

Next day, we rode out to the scene of battle and re-enacted it for the benefit of our comrades, then butchered our kill, retrieving a roast and as much of the fat as we could carry. The bear was as fat as butter, in preparation for his hibernation, and it was not all wasted, for besides what we took, a forest ranger happening by our camp was given as much as he wanted. Bear grease is greatly valued in country districts.

More hunting being for the moment an anticlimax, we packed up and plowed through the snow of Two Ocean Pass again, to rejoin the left-behinds. We saw a few belated elk, but were halfhearted about hunting them; and a bull moose,—who, filled with curiosity, followed in our trail so close that for a minute we thought we had acquired an extra pack horse,—was simply shooed away.

We found our friends rather disgruntled, as well they might be, at being left alone in a gameless country, but it was no concern of Pat's and mine. That Sunday we feasted on roast grizzly and "fixins," the joint efforts of the cooks of the party, which, with a nip or two from the diminishing contents of the whiskey keg, somewhat restored equanimity.

Then, Pat and I said good-bye to the Codyites, and with our own outfit bucked the deepening snowdrifts once more over Two Ocean, to camp on Soda Creek.[8]

Elk were all around us but we did not delay, for night in our stoveless tent reminded us that winter was near. We pushed on, over the rough bridge spanning the Buffalo, to Black Rock Creek.[9] Ahead of us in the snow were the tracks of many elk. At noon we overtook the rear guard, a band of young bulls, and from them Sherwin replenished our meat supply by a neat shot.

At sunset we heard bugling and saw a few in sight to our right, and, hoping for a better head than I had, I rode over to investigate. Leaving my horse, I crawled through a tongue of timber reaching out into a broad meadow, and near dusk found myself in the midst of a band of two hundred or so. The bulls were calling their challenges, and among them I heard the deep bellow which could only come from an old-timer. I saw a bull raise his head to bugle, and heard the deep note. Although nearly two hundred yards away, I knew it was my only chance, and fired. At once, the whole band streamed past me, my bull among them, out of sight in the timber. Then into the open, alone, came an enormous bull, no doubt the real author of the deep bugle. He stopped to stare, then hurried out of sight, untouched by my long-range bullet fired without aid of sights. I followed the herd over the ridge and found the bull I had shot at first, dead, much to my chagrin, for he was not worth the killing, a case of mistaken identity.

My horse had strayed, and when found, stampeded at my approach. When I at last caught him after a chase through snow-laden brush, a cold, tired and disgusted hunter arrived in camp. That night was a bad one. A hastily chosen camp site exposed us to wind and drifting snow. Smoke filled the tents, and blankets felt as thin as cotton. Uncomfortable as we were, it was harder for the horses. Too cold to try for grass, they stood humped up all night, and morning gave them no relief.

We packed and saddled, breakfastless rode on solid ice over the creek which we had waded the night before, and though the bugles of elk resounded in the cold morning air, they were not a siren's call for us. We busked the snowdrifts of the continental divide to the shelter of heavy spruce timber, then down to bare ground at last on the head

[8] Soda Creek is in the northwestern section of Yellowstone National Park.
[9] Black Rock Creek is only a few miles east of Soda Creek.

of Wind River, where we pitched our tent, rested and fed ourselves and our weary horses.

Our hunt was over. If my account of it sounds too much like the usual magazine hunting story, I am to blame for telling it in such detail, yet to me the scenes and events are so vivid that I feel they are worth recording. The perchance reader may skip them if he will.

As I get older, though I love the wilderness as much as ever, I feel less desire to hunt for the sake of killing, not that I ever was a game hog; but it does seem to me that we killed a lot of animals on that trip. However, game was plentiful then compared to now, and though we tried not to waste, we saw no reason, if a better head appeared than the one we had, not to get it and discard the old, or if our meat supply was tough or stale, to get a fresh supply, and take only the choicest parts at that. Under the circumstances, we couldn't burden our pack animals with too many game products.

Fences told us we were again entering civilization. At camp time we rode into the little town of Dubois,—one store, one saloon, and a house or two,—where we pitched our tent and corralled our horses for a feed of hay.

In the saloon, where we went to celebrate our return to the haunts of men, we found an army officer from the post at Fort Washakie, Lieutenant Enslow, 10th Cavalry, who with his colored striker had likewise been on a hunting trip. A very pleasant gentleman we found him, and, one thing leading to another, we soon established a stud poker game with him, the barkeeper, a ranchman, and a long-haired individual from the backwoods, while the Lieutenant's striker served liquid refreshments and attended to our wants betwixt his grinning absorption in the game.

The game broke up in the wee small hours, without much damage to anybody, because the total funds of the party were modest and circulated fairly evenly with the changes of fortune. At 5 A.M. our Lieutenant bade us farewell, begging us to look him up when we passed through, and vanished down the road, with Washakie, seventy miles distant, his avowed destination by night. Pat and I followed more leisurely, as parting glasses and songs seemed necessary to cement our friendship with the Duboisans, but we got away at last and jogged in a snowstorm for thirty-four miles to the J.K. ranch, and camped. On the fifth we resumed our march over barren, waterless plains, Shoshone Indian country, but fit only for prairie dogs. The only incidents I remember were the sight of a sleek army mule lying dead on the plain, with no mark upon him, whether driven too hard by our military friend in his effort to reach the post before his leave was up, we never knew; and the sudden appearance over our heads of hundreds of sage grouse on migration, their alternate quick wing beats and sailing giving a curious effect en masse. At some meal we had, an unkempt individual rode up, and, invited to join us, refused food but asked for coffee when we had finished. He filled the pot half full from our previous store, and made a kind of gruel of it, then boiled and ate it! Just a horseback tramp, he was.

At Washakie, a group of houses around a square, interesting because of their age, about 120 colored troops with their officers were stationed. We looked up our friend and found him in spick-and-span uniform, with which he seemed to have donned formal manners far different than the happy-go-lucky poker player's of a few days back. True, we were rather a tough-looking pair, in leather chaps and grease-stained hats, and perhaps he was to be excused for failing to invite us to meet his friends.

From here on, it was just a ride, through Lander, then having a building boom in anticipation of the coming railroad, then over a divide through hard-packed drifted snow to the group of little mining towns, Atlantic and South Pass, my old stamping ground of seven years ago. The latter looked hardly changed from the old days, but inquiries failed to locate any of my former friends. Populations shift rapidly in these outposts. At Pacific Springs, just as it was when Cherokee Dad drove us, everyone was gambling and well charged with red eye. Perhaps they were still playing for the loot won from Dad's party.

Again we made Washington's ranch in an uneventful ride, and on November eleventh rode into Rock Springs. We had sold and traded our best horses along the way, and arrived with a sorry outfit for which a livery man grudgingly paid us seventy-five dollars, enough, however, with what we had, for a luxurious Pullman ride to Denver.

One would think after six weeks of camping we would have had enough, but even that last two hundred miles across Wyoming in a gameless country hadn't tired us of the trail. For a long time I for one, who in addition had had a summer of life in the open in Utah before this trip, missed everything about it, the crisp morning air as we threw the tarp back and rubbed the sleep out of our eyes, the smoke of the campfire and the smell of good bacon and coffee, rounding up the horses for saddle and pack, and above all, the feeling of being on one's own, with a good and competent outfit, providing for ourselves and asking help from no one. I could have kept on forever, I thought, and resolved to get back to the wilderness as soon as possible. But it never happened again in just that way. It never does.

On return, I found the already recorded windfall in Esperanza stock, and feeling a rich man, it took little persuading on Halstead Lindsley's part, who painted in glowing colors the joys of a trip to Havana, to cause Sherwin and me to join him in a rail trip to New Orleans with the Cuban city our destination. We revelled in New Orleans oysters and Ramos gin fizzes a day or so, then, finding that Havana was under quarantine for yellow jack, took ship on the *Comus* for New York.

In Boston, I had my usual good time with family and friends, saw Yale beat Harvard according to custom, had a good cross-country ride or two with friends in Westwood, then began to think of getting back to serious work.

I guess I was a little talkative on this trip home about life in the West. Nowadays, plenty of Boston people know farther places than I knew, but that was before the days of dude ranches and expeditions to scale remote peaks. I was sitting at lunch in the Tennis and Racquet Club one day, and having told among other things of the killing of old Four Toes, said to the quiet and interested man beside me, "You really ought to take a trip out there sometime and try for a bear. It's the best sport of all." A grin spread around the table, and someone said, "Look at those heads on the wall." I looked; there was an immense Kodiak bear, half of him mounted as if springing out of the wall, and beside him were two grizzly heads, each as big as mine, looking like cubs beside the monster. Turning to my neighbor, I said, "You—?" "Yes, I killed them," said he. He was Jim Kidder, a mighty Nimrod. I have talked small about that bear ever since.

It is curious how in a man's life some trivial episode appears to mark a distinct crossroads, where if another road had been followed one's career might have taken a different course.

Lindsley and I sat in Rector's restaurant one day. He was going to Egypt; I to Colorado. He urged me to come with him, holding out as bait a camel ride over the Sahara, to take a look at the mines supposed to be King Solomon's. I was half minded to go, but conscience pricked. At last, I said, "We'll toss a coin. Heads, I go with you; tails, back to work." A double eagle spun upon the table. Tails came up.

I went to Denver, and there found a letter from a charming young lady whom I had met in Colorado Springs, inviting me to dinner and a dance at Glen Eyrie, General Palmer's castle. The lady's name was Gwendolen Young. We were married in June.

As this is the record of the highlights of my bachelor days, and as the life of a married man takes on a different pattern, this is the appropriate place to end this journal. If ever I complete the tale, it will have the advantage of supervision by my family, who have become increasingly a part of it, and will on that account need another cover.

XIII

High Society and War in Telluride

When I began my journal at the age of sixty, I made the remark that instead of enjoying the quiet of the farm I was still on the road and that home life seemed as far away as ever. I finished that journal by saying that as it ended with my marriage, a continuation would need another cover, so here it is.

Here I am, seventy-four years old, still in business but now more at home at the farm than I have been for the last few years, and at last I have the leisure to make a try at that other story. Let me say that as it is primarily for my own amusement, I am free from the inhibition caused by apprehension of others' reading and criticism; hence if anyone does read it, I make no apology for lack of incident, or of literary merit.

It is difficult to know how much to include or to exclude in a journal of this sort, and to record things chronologically sometimes befogs the story, so I'll avoid the diary form and as I can only become tedious to myself, I will set down the things that occur to me and linger in memory, trivial as they may be.

To go back a little,—in 1905 I was leasing in Cripple Creek, where, as I had a good working partner, I spent a good deal of my time exploring the surrounding country ahorse. I had a good pony, in fact, a succession of them, and I used occasionally to ride the forty miles to Colorado Springs, where my sister was then living, and there enjoyed the contrast of civilized life to that of a rough mining camp. Here I first met Gwendolen, I think at Lucy Hayes' house, and through our mutual love of horses we became well acquainted on many a ride together.[1] In January, after a return from an eastern

[1] Lucy Hayes married Gwendolen Livermore's brother, George Young. A granddaughter of Jefferson Davis, she died in Norwood, Colorado, in 1966.

trip, I found a note asking me to a dance at General Palmer's Glen Eyrie, and that party led later to our engagement.

I already had a trip to Mexico planned, and went there in search of a nonexistent mine. My wandering took me to Mexico City, and to various remote properties in that country and in Arizona, but the only things of interest to me on return to El Paso were, first, the receipt from the grinning hotel clerk of a stack of letters all addressed in Gwendolen's flowing hand, and, second, a trip in the desert to the Bullard Mine in Arizona. The mine was no good, but the desert at that time of year was a garden of flowers and teeming with bird life. Every cactus was in bloom, mockingbirds sang, rabbits scurried, and plumed quail piped from every arroyo. I couldn't resist a day or two with a shotgun and, I believe, I took a few quail home to Gwen.

Our wedding in June was quite a gala occasion. Father, Harry, and Arthur Rice came from the East, and Halstead and Buck Wells were among my ushers. Gwen had a bevy of fair bridesmaids, including Gladys Young.[2] After the ceremony Alan Arthur[3] tooled the wedding party across the mesa in his four-in-hand to the reception at Glen Eyrie, and we left in a blaze of glory.

We went on our wedding trip to New York, then—of all places— to Lake St. John in the Quebec north woods. That mosquito-y place offered few diversions, and if we had known that we were later to live five years in similar outposts of northern Canada, and had I not been so woods-minded, we might have gone abroad together. Although in following years Gwen went several times with family or friends, I never got to accompany her. It was a missed opportunity, and one which is still a matter of mild reproach.

The first job offered me on return was to install a prospecting crew in Union Park, Colorado,—a basin high in the mountains up the

[2] Gladys Young, a sister of Gwendolen Young Livermore, is currently living in New York City.

[3] The son of President Chester A. Arthur, Alan was born in New York City in 1864. He graduated from Princeton in 1885 and "spent a year at Columbia Law School to please his father." After Chester A. Arthur's death in 1886, Alan banished from his mind any thoughts of a legal career and fulfilled his desire to be an international playboy, finally settling in Colorado Springs in 1899. There he assumed the role of a society leader on his beautiful Trinchera ranch estate, until the coming of the World War I. The war altered his world, but it did little to change him. He died on July 18, 1937 (Marshall Sprague, Newport in the Rockies [Denver: Sage Books, 1961], pp. 234–235).

Taylor Fork of the Gunnison River. I took my bride by wagon up the rough bed of the Taylor, called by courtesy a road, to the park, where George Young[4] with a crew had preceded me. There was a one-room log cabin, a line camp of a couple of cowboys, who gallantly turned it over to us with the loan of their horses when needed. A bed of fir boughs, a rough table, and a cookstove were our only furniture. There Gwen was supposed to keep house and cook meals for four hungry men. We still tell the story of how, when we came in hungry, she greeted us with the announcement that she could think of nothing to cook but did boil some water! How she trustingly rode alone through twelve to fifteen miles of wilderness to Tin Cup,[5] the nearest settlement, to get meat—at my instigation—disdainful of possible peril; how, when bathing in the mountain stream, she heard a noise and looked up to see several interested mountain sheep grazing on the banks, all, as I look back on it, quite an experience for a young and trusting bride.

We left George and his men digging holes and made our way back, after a thirty-five-mile ride to Buena Vista to advise Wells by telephone that there was enough gold in the preliminary work to justify sending in the keystone drill. Parking Gwen at Glen Eyrie, George and I resuscitated the drill which had been left at Mancos, and set out from Pitkin under steam for the Park. It took us a week to do the fifty miles, up over a divide 12,500 feet high, delayed by a dished wheel, which we got repaired at the little town of Tin Cup, a broken bridge or two, sleeping where night overtook us and, at last, weary and disgusted, we made the Park. I left George in charge and, picking Gwen up at Colorado Springs, took up residence in Telluride, our home for the next four years.

Due to Hal Lindsley's and my friendship and mining connections, I had been offered, and bought, a quarter interest in the Lindsley King lease on the Smuggler. Arthur Townsend had bought the other

[4] George Young was a rancher in Norwood, Colorado.
[5] Tin Cup was first known to the mining world as Virginia City. It was designated Tin Cup in 1882, by which time it had a population of 6,000. Two years later, the population dropped to 400. At the turn of the century, Tin Cup took on the ghostly air of many another mining camp. Another boom hit Tin Cup in 1903, only to fade by 1917, when Tin Cup's main street again was empty (Muriel S. Wolle, *Stampede to Timberline* [Boulder: University of Colorado, 1949], pp. 180–187; Nolie Mumey, *History of Tin Cup, Colorado* [Boulder: Johnson Publishing Co., 1963]).

quarter, and the lease was thereafter known as Lindsley, Townsend and Livermore.

Life in Telluride was far from humdrum. The town was still on guard against the union subversives, and rumors, sometimes becoming facts, were in the air. We rented a house in town which was comfortable enough, except, for economy of heat, the only access to the bathroom was through the combination living and dining room. However, we managed social events well enough. Our maid of all work was unique. After much search we settled on a lady with a past, a graduate of the "row," who must have had attraction, as she subsequently married twice. At that time she was a Mrs. Wardlaw. She refused to work for anyone but Gwen, whose broad-minded attitude won her devotion. She was with us much of our stay in Telluride, an excellent cook and an exemplary character.

We were well mounted; I had several good horses and, later, bought a nice grey mare for Gwen whom we named Chispa. Everybody rode; the mine and the mill were miles apart and horseback was the most convenient way of travel. The livery stables flourished, for the travel between town and the various mines in the mountains by miners going up to work or coming down on holiday was constant. The rule was to turn the horse loose at the mine, when he would find his way home.

A great deal of the supplies were carried up by mule train. It was common to see a string of fifteen mules loaded with cases of dynamite or provisions, or dragging planks or timbers lashed on either side, their ends dragging on the ground. These trains gave ground for no one. Many a time our saddle horses danced precariously on the edge of a cliff, or, if we chose the inside, were squeezed against the rock by mules who objected to being crowded.

There was a gay crowd of eastern emigres,—Lindsley, Townsend, Pierre Barbey, Max Henderson Scott, a young Englishman, Barnett, another, and later, attracted by our example, Bill Clark, Billy Rogers, Bill Stickney, Bert Kruger, Sidney Fish, Oliver Filley, and others. Buck Wells was the kingpin, adored by all who had followed his leadership, and Grace and her four children had taken up residence in the company house at Pandora a year or two before.

The town was wide open. I believe there were thirty-odd saloons, in most of which a roulette wheel and faro game were established. The feeling among the mine operators was that the sooner the miners got broke, the sooner they went back to work. The "row," a street near the river and railroad track, was inevitable, a feature of all mining camps of the day.

The New Sheridan Hotel was the mecca of visiting travellers and of the elite of the mining men and cattlemen, who liked a slightly better brand of liquor or cigars at the bar presided over by Jimmy Allen, hail fellow well met.

Telluride being on the verge of the cattle country, there were always a few cowboys in town, their ponies standing patiently at the hitching rails outside the saloons, and after roundup in the fall when cattle were shipped, the town took on a decidedly frontier aspect. Usually a well-patronized poker game went on in the Sheridan, between cowmen with well-filled pockets and the local talent. One such, I remember, lasted twenty hours, with lunch and other refreshments served on the table.

The feminine contingent was not lacking. Gladys Young was a fairly frequent visitor, Lucy Hayes, who afterward married George Young and became a fixture, Charlotte Hemenway, Frances Curtis, Bill Clark's sister and mother, Bill Roger's wife Grace (née Chapin), May Lindsley, then a dark beauty, and others. My brother, Tom,[6] attracted by my success in leasing, came up from Mexico with Sibbel and, eventually, started the ill-omened Lewis Mine which sadly cramped his and father's finances.

Lorna Stimson was another who visited us and who fell victim to the life, the beauty of the mountains, and to Hal, picturesque in his wide Stetson, his high-heeled boots, and his handsome steed. The story is—I don't know how true—that when Hal came east for their wedding, his appearance, garbed in conventional clothes, his too small derby hat perched on his shock of Japanesy black hair, was such a shock that Lorna thought seriously of calling off the engagement.

Our arrangement for running the lease was for each partner to take his turn at management, which left the unoccupied ones to use their leisure as they pleased, whether for play or work. This gave me time to do examination work for others or to take a holiday with Gwen. It worked fairly well for the lease until during Townsend's incumbency, some time later, things went very badly and Hal and I

[6] Thomas Livermore later forsook a mining career to become a citrus grower in Florida. He was killed in an automobile accident in 1935.

prevailed upon him to retire and leave the running of affairs to us. Halstead, always the inventive one, had the idea that the old stope fills of the Union, left by the old-timers, who shipped only what would later be considered high-grade ore, could be profitably mined again. These stope fills were cemented by time into a fairly solid mass. After many experiments we found that by mining small blocks as quickly as possible before exposure to air loosened up the mass, then flooring over the stope for the next level above, we could take out nearly everything; and by sorting out the waste outside, could ship a payable ore to the mill. I got George Young, who, though lacking in mining experience, had plenty of calm, common sense, to act as my superintendent. This operation was profitable for the two years that the ore lasted, and added greatly to our tonnage. I believe that at one short period we were running 120 stamps in the mill to the company's 20.

Our first winter passed pleasantly enough, mostly occupied by the routine of riding up to the mine each morning and down each night, with occasional rides down the valley to escape from the snow which, the higher you went, the deeper it became; a feast at Christmas replete with song and story; and a New Year's open house to which the townspeople were invited, and at which a flowing punch bowl with a substantial "stick" contributed to the gaiety. In fact, I think that some of the ladies were unaware of the stimulant and the cause of their consequent high spirits. The local paper reported the event as one of great éclat and mentioned "the light of innumerable candles" as part of the decor. I think there were four.

In January my term was up and Gwen and I went east for the round of Boston and New York pleasure. My Boston friends were no doubt intrigued to meet my Colorado bride, and found her acceptable, judging by the number of invitations we had. We had a slight suspicion that they had expected someone with feathers in her hair.

We found plenty to do until my return in April, when I went to Telluride to resume my term at the lease, leaving Gwen in Colorado Springs. My temporary bachelor's life was marked by several rides in the surrounding mountains in company with Hal, Bill Clark and others. We often rode over deep drifted snow which at this time of year was compact enough to bear our horses.

One of our enterprises was the so-called Telluride Land & Lumber Company, which owned a considerable tract of forest land on the mesa fronting the San Miguel River. Here was a noble stand of yellow pine (ponderosa), which was logged and transported by aerial tram to the sawmill in the river valley. It was a lovely place; open prairie on the higher slopes and parklike groves of enormous pine below, through which we rode rejoicing. Small game of various sorts, blue grouse, sharptails, ducks in the sloughs, and occasional deer, were fairly plentiful and graced our fare. Once I flushed a cinnamon bear and once we saw fresh lion tracks.

We reserved a sizable grove of pine near a clear spring, for a cabin, and engaged George Young to serve as architect and builder. He did a wonderful job. It was made of logs, with a huge fireplace and room for all. In it we camped, cooked and slept whenever the spirit moved us to ride the thirty miles from Telluride.

The lease was doing well; success was just around the corner. As I look back on it, it was one of the happiest periods of our earlier married life. "Treetops," as we called the ranch, was headquarters for many trips in later years. An expedition made up of the men, our wives, sisters and feminine friends, mounted on good horses, equipped with pack horses, made a cavalcade worth watching. This was on the edge of the cattle country, as yet untrammeled by fence or cultivation. One such trip was made to the roundup where all the beef of the range was being gathered for branding and sorting. A chuck wagon gave forth its plentiful and delicious fare, and the dust and bawling of the calf roping served as chorus. These cowboys were quite as skillful with a rope as any of my Montana friends.

That year (1907) we moved to a little company house at Pandora, at the mills, two miles above town, altitude 9,000 feet. The roar of sixty stamps was in our ears, but so accustomed to it did we become that we waked only when for some reason the stamps stopped. This house was formerly one of three almost in line with the Ajax snow slide. One had been destroyed years before, but the other two were thought to be safe. Years later the Ajax ran again and wiped these houses out, killing the housewife of one and burying the wife and child of the bookkeeper, who lived in the other one then, under tons of snow. His wife was killed but the child was miraculously rescued.

Avalanches were a constant danger in that country. Some years before, over twenty men were killed in the Liberty Bell slide, and hardly a year passed but one or two unlucky ones were caught. In a spring thaw an enormous boulder plunged down the mountain into the front lawn of Wells' house, and on the next bound went through the roof of the cyanide plant. No casualties, however.

Gwendolen's and my riding activities were curtailed by the expected arrival of an infant, but I am afraid in our ignorance we did far too many strenuous things. For instance, we took a couple of trips to the "lower country" by wagon over rough roads which, undoubtedly, were not good for an expectant mother. In one such, to "Treetops," we dished a wheel and had to hurriedly disembark. George and I rigged a stout pole in place of the wheel, and with it we got back to Placerville, where we found a replacement, but everyone was pretty well tired out when we got back to Telluride.

It became evident pretty soon that all was not going well, and poor Gwen spent many a lonely and depressing day at the little Pandora house while I made my daily trip to the mine. The inevitable happened, a premature birth, an ordeal from which Gwen escaped without apparent injury, but life in Telluride had lost its savor, and as soon as possible she went to the safe haven of Colorado Springs for recuperation. I joined her in October en route for Harry's wedding in Boston, which event we attended and enjoyed to the full. Then we spent a few days with Mark and Gladys Hopkins at Dublin, New Hampshire, chasing the wily partridge.

At General Palmer's insistence (and by his financial aid) Gwen went to Europe, alone, to my sorrow, for further recuperation and to chaperone Gladys, and I went back to the mines somewhat depressed. On Gwen's return it was apparent that still all was not well, and an operation at Colorado Springs was decided upon, from which, thank the Lord, she came out well. However, I shall not forget my apprehension as I wandered in the fields outside the hospital, comforted by Grace. I can still hear the meadow larks singing "cheer up; be not downhearted," and sure enough, they were right.

The winter of 1908 saw us installed at Pandora with "Annie," a faithful but decidedly sketchy importation from Ireland by way of Denver, as factotum. We built, with George's aid, a new barn for our horses and thereafter fed and curried them ourselves, a real pleasure. One importation, an important one, was "Battle," a great Dane and a former gift from General Palmer, who owned and bred that type of dog, weight about 150 pounds, and in spite of his formidable appearance, as gentle as a kitten. He accompanied us on our rides everywhere. His courage was not of the highest order. One time when I left Gwen sketching while I rode off on some errand, I returned to find her at bay on a stump, surrounded by curious range cattle, exclaiming "I don't like cows!" Binkie, who should have been a protector, had high-tailed it to distant parts. In his puppyhood Binkie had been allowed to curl up in a chair, and in his age he still believed he could use that method of rest. I can still see him, his hind quarters only perched in our most comfortable chair, while his front legs sprawled on the floor. In our rather frequent travels by train, he was consigned to the baggage car, where the agent gave him a wide berth. He was quite a problem, but it was a case of "love me, love my dog."

Dogs being a necessity to us, I acquired several successive setters, my favorite breed, and tried to school them on grouse—without much success. I even had my favorite, Dazzle, shipped on from Boston. That was a mistake, as Dazzle, already old, found the change of climate too much, and died after a lingering illness. Our final setter was Tamarack, whom Gwen bought as a puppy. Tammy was a real character, too headstrong to make a good shooting dog, but a loyal friend for many years wherever we went in Colorado, Dedham, Canada, and later at the farm in Boxford.

Chispa, whom I have mentioned, was a grey mare I bought for Gwen. She was a nice-looking, able horse, but was rather lazy. I used to say that she and her owner were alike in one thing—they were both suspicious of muddy crossings, and would shy off simultaneously. I remember my feelings were quite hurt at their lack of confidence in my pilotage.

I had several horses. Snip, a good-looking roan who was an incurable shier; Jimmy Britt, a fine-looking ex-polo pony; and finally Jamaica, whom I bought at the roundup. Jamaica was a showy roan whose best gait was a swinging trot, but rough and bouncy in every gait. He had wonderful endurance and was gentle enough for a fairly good rider, and was my best horse for all our stay. With him I certainly had plenty of exercise for the liver.

The summer of 1908 was the high point of our good times in Telluride. Many of the events I have mentioned occurred this season. We had a succession of visitors, some to remain, others to sample the life and go home. Riding by one and all was the pastime. Horses and mules were the only means of transportation. I think the first automobile appeared in the streets of Telluride about then, but except for a motor stage, which ran at the risk of many breakdowns over the winding and precipitous San Miguel canyon between Placerville and Norwood, there were few followers in our day.

Winter set in early. October saw the first real snow, which rarely left until April, getting deeper each month. Outdoor activities were limited, but the roads were kept open and often we travelled by sleigh or wagon down the valley, sometimes to get a last glimpse of the sun which had already set at Pandora, and once or twice on a winter trip to "Treetops," where we enjoyed a roaring fire, camp cooking, and such sport as a foot or two of snow permitted.

To preface the following episode, one of the aftermaths of the Western Federation of Miners' strike was the trial of Orchard and Adams for the murder of Governor Steunenburg in Idaho. Orchard was convicted but Adams not convicted, I believe, because he repudiated his confession. At any rate, he was to be tried for crimes committed in Telluride and was in jail there at this time. The story was that the union had promised to get rid of the witnesses against him, and Wells was one of these.

To digress a little, Adams was an ex-cowboy, and except for being a paid murderer, a rather likable sort of chap. In fact, he differed little from the hired bravos on our side, except that he happened to choose the other tack. One time he offered to show Wells, the sheriff and others some of the scenes of various unexplained crimes, if only they would let him "fork a horse" again. The offer was accepted, a horse was provided, and the group escorted Adams around the district. He showed where the shotgun which killed Collins was tossed after the murder, then offered to find the remains of Barney, a Smuggler-Union shift boss who mysteriously disappeared during the strike.

It seemed that Barney, after an evening in town, was saddling up in the livery stable to return, when a group of union men surrounded him and someone shot, but only wounded him. They then mounted him and rode him a little way down the valley, all the time reassuring him that they intended him no further harm. Then they shot and killed him and tossed his body in some bushes. Later, the tale went, the odor of death being pretty strong, they moved the body up the

Alta road and put it in heavy timber some distance away. Adams pointed out the locality, but no body was found. The crowd was getting rather skeptical, when it was suggested that the snow might have moved it farther down the steep slope. Sure enough, the bones scattered by coyotes were found, and poor Barney's skull, still partly covered by his red hair, was sure proof of his identity. The pitiful remains were exhibited in a Telluride store window to convince the union sympathizers that the murder had been committed, but I believe they only commented that the "mine owners had done it to discredit the Union,"—the usual union explanation.

On March 30-I was awakened by a mill foreman saying I was wanted at once, that the "General" had been blown up. I hurried a coonskin coat over my pyjamas, pulled on my boots and rushed over to the residence, expecting to find Wells in pieces. To my relief, he was sitting at his desk, his face covered with blood from numerous scratches, but calm and otherwise unhurt. The attempted assassin had placed a time bomb under his bed, but as Wells slept on a sleeping porch, the force of the explosion was downward, and the flimsy roof of the projecting room below took the force of the explosion. The further fact that his mattress and coverings were thick saved him from the direct force. He, bed and all, was lifted through the window beside him, which accounted for his numerous cuts and scratches. While we who had gathered were discussing matters, horsemen were heard outside the door, and cries of "Come on boys, let's get Adams! No more good men to be killed because of him!"

In a moment my horse was saddled, and we all set off on a gallop for the jail. We were let into the sheriff's anteroom, which was separated by ceiling-length iron bars from the prisoner's section. It looked pretty bad for Adams. Beside us deputies, a crowd of townspeople were gathering, and the atmosphere was decidedly ominous. Adams was cool, smoking a cigarette and exchanging remarks with his visitors, though he well knew what might be coming. Just then up drove Wells, who, realizing what was up, had a buggy harnessed and sped to town. "Boys," said he, "if you have anything in mind for Adams, don't do it. If we get one legal conviction, our troubles with the union will be over. If we make one illegal move, our name is mud!"

There was nothing to do but consent, so the would-be posse of lynchers turned into a bodyguard for Adams until the excitement simmered down. I wish I could say that Adams was later convicted, but he was not. A clever lawyer got a change of venue to an agricultural county, whose jury wanted none of our troubles and released him. I don't know what became of him,—but that was the nearest I came to being one of a lynching party.

My term at the lease being over, Gwen and I decided on a respite from the San Juan winter, and sought warmer climates. We journeyed by train to New Orleans, took in the various sights, the Vieux Carré, Lake Pontchartrain, enjoyed the balmy air and famous food, then went on to Florida through monotonous pine woods, every tree deeply scarred by the turpentine gatherers. We landed in St. Augustine, rather disappointed at its mundane appearance where we had expected all the elegance of a Riviera; and thence went by Flaglers railroad to its terminus at Knights Key. We did get rather a kick out of going to sea by rail. We kept on by steamer, a rough passage across the strait to Havana, and there found ourselves at last in tropical air. After a few days of exotic sights and delicious viands, we decided to embark for home.

While circling by boat the remains of the *Maine* of '98 memory, we met, in another boat, Lorna Stimson, whom we were glad to join on the excursion steamer *Oceana*, and had a most pleasant voyage north. The events of the passage were a brief stop at Nassau, with which we were favorably impressed, and a stop at Hampton Roads just in time to see Roosevelt's fleet steam in from their advertising trip around the world.

We spent a few weeks among familiar scenes in the East, making 34 Alveston Street, Jamaica Plain, our headquarters, using, as a matter of course, father's boundless hospitality, his horses, and his car—at that time, as I remember it, a one-cylinder Oldsmobile in which I drove my required two hundred miles under tutelage in order to get my driving license.

That car was quite a character. The engine was in the rear, and various chains, sprockets, and gadgets transmitted power where needed. It usually went quite satisfactorily on the level, but the radiator tended to boil and the engine to fail halfway up the moderate hills of the neighborhood, when a neighboring brook was called upon for cooling purposes. At times a clank in the machinery was heard and all forward motion stopped. We learned to go back on the road a little and pick up the truant part, fit it in where it seemed to go, and usually went on our way. The car finally met its end by turning over on a curve, with my brother Harry driving, and burning up. Harry, who had dined rather well, had some difficulty extricating himself, but did so unscathed.

We spent a good deal of our time that spring in driving about the country looking for my obsession, a country place that I could eventually call my own when my fortune was made and I could settle down to the life of a country gentleman. I don't know what Gwen, somewhat more urban-minded, thought of it, but she loyally seconded my efforts. I gave myself five years in which to make that fortune! We finally found a place in Norfolk, a couple of nice old farmhouses on a clear lake with pine-clad shores.

The trouble was that the best house and half the land was owned by an obdurate Irishman who refused to sell. I ended by buying the smaller house and gave the old lady who lived there, Mrs. Kingsbury, a life tenancy. However, we never took possession. Mining life was too exacting, and eventually we sold at small loss.

When it came time to return to Telluride, Dr. Broughton vetoed Gwen's going, as a baby was under way and premonitory symptoms were not too favorable. In view of the last episode, the verdict was for her to stay quietly in Jamaica Plain. So, reluctantly on both our parts, we separated, I with lone months in Telluride as a prospect.

Resumption of a bachelor's life lacked a little of the savor of old, but the routine of looking after a couple of mines soon occupied my thoughts, and I can't say that I acted the disconsolate husband too much. There was simply too much activity around Telluride to allow moping. "Scotty" (Max Henderson Scott), who by this time had become a right-hand man for the lease; my brother Tom; Malta, a Dane bitch, another gift of General Palmer's; and I took up residence in the Pandora house, imported Annie Shea from Denver, and settled down.

The routine of mining work was varied by the never uninteresting events of a flush mining town, still with its rough edges not too smoothed. I remember one "tin horn," named Munn, semi-cowhand, semi-gambler, who killed the sheriff's deputy who was trying to disarm him, and took to the brush. The deputy, Art Goeglein, was universally liked, and a posse formed and hunted the killer far and wide, but without avail. Finally the tin horn came in and gave himself up, saying, "I got tired of waiting for you fellows to catch me, and fed up with living like a chipmunk in the brush." I believe he was acquitted, as usual, on the grounds of self-defense.

Fourth of July was an occasion for much celebration. The usual events were, among others, a drilling contest by teams of double hand drillers who drilled an incredible number of inches through tough granite;[7] a mule-packing contest where the packers each led his string of mules—fifteen in number—while his partner loaded each mule with three sacks of heavy mill concentrates, lashed them securely, and tied the loaded mule to the one ahead. I believe the total time was in four or five minutes! Every bar was open and every gambling device did a rousing business. Miners, cowboys, and citizens "turned her loose" for that day.

This particular year there was a rodeo at the San Miguel race track for racing, cattle roping, and buck jumping. The town was still a bit jumpy from recent strike events, and several of the "red necks" were known to have returned. Wells cautioned all well-wishers, in particular the special deputies of whom I was one, to display no weapons and to start no trouble. Nevertheless, most of us tucked away a six gun just in case. My particular safeguard was a colt automatic, which I thrust into the waistband of my trousers.

While watching the contest in the thick of the crowd I heard a shot and felt a simultaneous blow on my leg. Not until a moment had passed did I realize that my automatic had gone off and that the bullet had ploughed through my leg. I tried to bluff it out in spite of the blood running into my high boot, but the squeals of the man standing next to me, through the calf of whose leg the bullet had also passed, attracted the attention of the crowd, and I was hustled off the field by my friends before rumors should develop into a riot. In the

---

[7] Hard-rock drilling contests were a favorite sport in mining camps throughout the West, especially on the Fourth of July. Contestants would come out of the mines two weeks before the event to go into training for the event. Specific rules were laid down fifteen minutes before the contest began. Teams of two men each would hammer at a block of granite with hand-sharpened drills. When one drill became dull, the men would exchange places, switching drills, and seldom miss a blow. One man would turn the drill after his partner had delivered the blow. They alternated back and forth until a whistle sounded, ending the session. The depths of the holes in the blocks of granite would then be measured, and the team drilling the deepest crevice would be declared the winner. Occasionally, a team would drill thirty-two to thirty-five inches in hard-rock granite in fifteen minutes (Victor I. Nixon, "Hard Rock Drilling Contests in Colorado," *Colorado Magazine*, XI [May 1934], pp. 81–85).

hospital I was told of the other man's injury, and was gently re-proached by Wells for being the cause of the commotion. In the end, a couple of weeks' inactivity and a cop to my victim in the shape of doctors' expenses, ended the episode, but not my chagrin. I guess some pre-sports potations had something to do with the matter.

It was shortly before this episode that there was quite a lot of excitement caused by Wells' announcement to a select few that the long-delayed San Miguel Development Company ditch[8] was going through to the Dry Creek Basin, vacant semi-arid land, about fifty miles down river, hitherto the prized winter range of lower-country cattle.

Several of us set out by wagon and horse, camped at Coventry on the ditch, and next day, via the "long draw," made Lavender's[9] Spectacle Camp in the Basin. This land was a flat, sagebrush-covered prairie, deep soil waiting only for water, intersected by deep arroyos made by the wandering creek, and surrounded by low mountains backed by the Lone Cone, a lovely pyramid on the south, and the distant La Sals in Utah on the west, a really wild and lovely sight.

After camp breakfast all hands were making ready to ride, looking for likely places to file, when a rider appeared who asked for me. He told me that Holland, the Smuggler Mine superintendent, taking advantage of my absence, was mining out the floor under our lowest level at the Sheridan, which, if carried out, would stop our operation, —and that this was with Wells' permission. After confirmation, after a reproachful conversation with him, I gave up plans for filing and mounted my steed for Telluride. I think this was the beginning of my distrust of Wells, at what I considered a treacherous act, and I am sorry to say that time did not completely heal the hurt.

I straightened things out in Telluride as best I could, then, with a hired team and with Bert Kruger for company, drove back to Dry Creek in hopes that some good land was left to file on.

Our trip was a hard and disappointing one. We heard from some outgoing filers that Wells' party had gone to the Fall Camp at the

[8] An extensive irrigation, reclamation, and hydroelectric scheme, the San Miguel Development Company never became much more than one of Bulkeley Wells' dreams.

[9] Edward Lavender, a long-time resident of Telluride, whose son David is a well-known western historian has written much on the West. His two best-known books on Colorado are *One Man's West* (New York: Doubleday, 1943) and *The Big Divide* (New York: Doubleday, 1948).

head of the Basin, and as we were already well on our way to the Spectacle Camp, we determined to cut across country to save retracing the miles. It proved a grievous mistake, as the sagebrush plain, deceptively smooth to the eye, was cut up by numerous arroyos which necessitated long detours. When we finally came to the main creek and found it at the bottom of a thirty-foot cut, we were really discouraged. Rather than give up, we unhitched the team and, belaying a rope around a stout sage bush, lowered the buggy to the bottom. Then, with the aid of the horses, we hauled it up a cow track on the other side. What with that, and several other episodes, we arrived at the Fall Camp thoroughly tired out, only to find Wells' party's tracks pointing out over the divide. There was nothing to do but keep moving, so late at night we arrived at Norwood, where we stabled our exhausted team and tumbled into bed.

I made Telluride the next day, but only by leaving my worn-out team at a ranch and being ferried in by the rancher.

That summer was varied by a trip east to see Gwen. I found her well but lonely. I installed her in a comfortable house in Marblehead Neck, and, after a few days sniffing salt water, bade her adieu and returned to Telluride. As the lease was running fairly smoothly, I found plenty of time to range the surrounding country a-horseback, mainly to "Treetops," in which my incorrigible instinct for land gave increasing interest. I guess my enthusiasm infected Hal, Bill Clark and the other proprietors to develop the place as a going ranch. We fenced a considerable acreage, planted crops suitable to a semi-arid climate, and invested in a few mares, most of them pretty aged, with the object of breeding them and starting a horse ranch.

As I see it now, the land was not suitable for steady crops—too high and too dry—and without competent and continuous management, affairs were bound to go haywire. There was a fine, clear spring in the "draw" near the cabin, and by a simple dam we created a fair-sized pond. Hal and I surveyed a reservoir site, but it was never completed, and, as I realize now, being dependent on rainfall, would never have furnished a reliable irrigation source. Another handicap was that this land was in the recently created Forest Reserve and, outside of our fee, was not open to filing. Without ample pasture, our proposed horse herd would have been distinctly limited.

We happened to live when, under the leadership of Teddy Roosevelt and Gifford Pinchot, the public was beginning to realize that the people as a whole were the owners of unlocated lands, and that individuals without title had no right to cut and use lumber and to graze stock on the public domain without compensation and regulation. Long and loud were the howls of the local stock and timber magnates, who had for years been cutting and grazing, with only a pretense of land tenure; and when forest supervisors invaded the country and began to assess the extent of their past trespass, there was nearly a revolution. One prominent lumberman, I remember, had logged for years with only a filing on one or two claims as a mill and camp site.

Fortunately, we had only gone over our bounds to a small extent, perhaps because we hadn't time to progress further, and our over-cutting was construed as an unintentional trespass, at small cost to us. In the end we were reconciled to using the place as a picnic ground, and brief as that use was, it gave us lasting pleasure.

The mesa country below Telluride was wild and beautiful still. From the breaks and cliffs overhanging the valley of the San Miguel, great open forests of yellow mountain pine reached toward the higher lands above, on the prairies of which grew tall oak brush and sage. All supported a luxuriant stand of the best graze, and every-where the rides were superlative. Higher still the foothills began, and the moister climate produced clear streams, dense spruce forests and hidden lakes. Here deer abounded, trout swam in beaver-dammed streams, and occasional bear left their tracks on the loamy banks.

One trip I remember was to Woods Lake, a beautiful sheet of water partly pre-empted by Telluride people who had built a log cabin or two. In one of these Buck and Grace, together with Bulkeley, Jr., and a trio of Welsh terriers, had taken residence. I joined them for a time and enjoyed to the full fishing the clear waters for the sizable but wary rainbows, seeing the beaver splash around their comical huts, and eating camp fare with vim.

One day Buck and I rode the long, winding Navajo Trail, the old trail which the Indian hunting parties formerly used, and on the divide gazed down upon the limpid waters of Navajo Lake and a wonderful wilderness which I mentally marked for future camping expeditions. I remember when the three Welsh terriers had en-countered a porcupine and came to camp bristling with quills, we spent hours extracting them from nose and throat. But I am afraid the dogs didn't profit by the lesson, as they tackled the next one they saw, with equal results.

This little outing was the more pleasant to me in my devotion to both the Wellses, as I thought it marked the end of their growing estrangement. But it was only temporary, I am sorry to say. Perhaps an accident to Bulkeley, Jr., who was thrown from a horse and received a serious skull fracture, from which he was recuperating, brought them together for the time.

In my loneliness for Gwen, I found distraction in these frequent trips away from the mines and sometimes, perhaps to accentuate my loneliness, preferred no company. One incident I remember, when I left my companion at "Treetops" and rode down the Beaver and across the San Miguel to the mesas on the other side, I saw game but could not connect until on the canyon rim near Placerville I ran into no less than three wildcats who allowed me a shot. My victim fell over the steep rim and I rushed to retrieve him, but could find no sign of him. While I was looking I heard a "miaow" and looked up to see just above me one of the other cats, apparently looking for his mate. He was so close that I tried to crease him with the rifle but evidently drew too fine a sight, and the cat vanished. I had never seen wildcats so unconcerned with the hunter before, and can only conclude that there was a love interest which momentarily subdued the natural caution. I came home wildcatless, as the one I hit evidently had crawled off to hide.

In August I heard from one of my best friends, Lanny Jay, that he was coming west to visit me. That summer was one of the rainiest I remember, even for the San Juan country where summer and fall rains were the usual thing. True to form, his train was stalled for thirty hours in a washout, and a somewhat disillusioned visitor arrived.

Hardly giving Lanny time for a breather, we went to "Treetops" as a start for a camping trip. After a night's rest we set out, each on a good horse, with Brownie, our faithful pack horse, and Malta, the great Dane, for company, on a camping trip—with Navajo Basin our destination. We made the Basin all right, but I found the spot which looked so attractive from the divide was on the rim of a canyon and unsuitable for a camp, so we fell back to a more level place and made down our bed in the open. In the night we were awakened by a downpour, and pools of water collected in our bed. Wet and miserable, we rigged up a makeshift tent from the tarpaulin and spent the rest of the night in comparative dryness—if not comfort. In after years

Lanny told that anecdote as an instance of wilderness resourcefulness, which to a city man seemed wonderful.

We were rewarded next day by a bright day, Colorado weather at its best, and rode through open forest to the top of the divide and one of the many fine views of this country; which inspired Lanny to remark that it was as fine as anything he had seen in the Alps.

Upon return to Telluride I resumed my round of the mines, and for a bit of local color took Lanny up on the tram and through the lengthy stopes, in places rather like holes through which we must crawl. It was the daily round for me, and I was much surprised when I asked Lanny, after return, how he liked it and he replied, "Bobby, I think it was terrible! How you live such a life in constant peril of burial alive, to say nothing of the danger of aerial transportation, is beyond me."

I remember on return from the camping trip Lanny took the first hot bath by courtesy, and then drew another and took that. He wasn't aware of our limited supply of hot water, and I got a cold one.

I returned to Boston in September and joined Gwen at Marblehead Neck, enjoying with her the few days of summer left. I invested in a Buick, my first-owned motor car, which was our joy and pride. It was a two-seater, no doors or top, acetylene lights, and an outside gear shift. I don't think there was a windshield, which was extra. However, it was a good little car, a big improvement over the sprocket-drive Oldsmobile, and had the merit of getting us to our destination and back in spite of occasional punctures. With it we toured the countryside far and wide.

In October our first child arrived, Bobby,[10] henceforth the pivot of our family life. While his mother was recuperating, I occupied myself with an examination in North Carolina near a little foothills town called Rutherfordton, for placer gold. There was a surprising amount of gold in the bedrock sand, but the area was small and an impervious layer of clay offered too great difficulties for dredging. This was hillbilly country. The natives, tall, rawboned Americans, were glad to work for seventy-five cents a day—top wages for

[10] Robert Livermore, Jr., became a founder and vice-president of the realty firm of Hunneman and Company, of Boston. He was an accomplished skier, and an account of his participation in the 1936 Olympics is in Robert Livermore, Jr., "Notes on Olympic Skiing: 1936," *Atlantic Monthly*, CLVII (May 1936), 617–622.

them—and sank pits to bedrock like magic. Too bad I had to disappoint them.

Eventually, my increased family being judged fit to travel, we set forth for Telluride. It was quite an expedition; a nurse, the faithful Florence, accompanied us. Besides the bulky baggage, an icebox containing milk, and all the paraphernalia for preparing a formula, were a part. At Chicago a representative of the Gordon Milk Company met us with a fresh supply, and finally we arrived at Colorado Springs without casualty. I went to Telluride alone to scout for quarters, then rejoined Gwen at Colorado City for the trip in. When we left for Telluride I well remember, in addition to "Battle," the accumulation of impediments piled on the platform. When the Pullman conductor saw it, he said, "This ain't an express car!" In desperation I remarked, "Conductor, aren't you married? Have a heart!" So at last we got to Telluride.

Houses were unobtainable, as Telluride was having rather a boom, but finally at great expense we bought the so-called Tallman house, a pleasant location with a lawn,—our dwelling for all our remaining stay. The faithful Mrs. Wardlaw, now Mrs. Schoolcraft, was installed as factotum. Winter was deep on the land, but we were so busy getting established in our new house that wintry blasts disturbed us not. Premonitions of petering out of the lease and our other mining activities were plenty, but in some ways the year was fuller of activity, both business and pleasure, than any that went before. Of course "Treetops" was the focus of our many trips. Everybody, whether horse-minded or not, was well mounted and thought nothing of the thirty-mile trip down the San Miguel and up Saltado Creek to the mesa.

Many were the pleasant, sometimes amusing, sometimes a bit worrisome, incidents of our expeditions. I remember one trivial bit of local color when a party of us had stopped at the hamlet of Saw Pit to ransack the one store for a bite to eat. One of the girls espied some tins of anchovy paste on the top shelf, and with visions of hors d'oeuvres in the future, demanded some. "I'll sell ye some," said the proprietor, "but I warn ye they won't stick nothin'. I've tried 'em."

Another time Hal and I went alone to the cabin, first each taking the precaution to tie a nice thick steak to our saddles. I had one of my various setters with me. Arrived at the ranch, we laid our steaks side by side on the kitchen table, and went to get some water. On return I was greeted by Hal, who returned first, "Too bad, Bob, your dog ate your steak!" It took me a few moments to realize that it was a mutual disaster.

I usually managed to keep the larder supplied with game from the rather numerous coveys of sharptail grouse, prairie chickens or occasional blue grouse, and from migrating ducks who dropped in at the numerous prairie ponds. Once we sent out our Japanese cook, Yeh, with instructions to get some venison, and jokingly asked him on return where was the deer. "He hanging on the gate," said Yeh, and sure enough, we found a nice spikehorn, neatly gutted and drilled through the eye, hanging up on the breeze. Once Buck Wells rode in at dusk, refused to enter the house to sleep but camped outside and produced deer meat from his pack, from which he prepared a delicious soup and roast. I always thought Buck had a little Indian, as well as epicure, in him.

One horseback episode was not so pleasant. Oliver Filley and Sidney Fish, two of our eastern visitors, both expert horsemen, took Bill Stickney on a tour of Telluride, with the usual accompaniment of numerous libations and visits to places of joy. Bill was totally unused to horses. In galloping around a corner, he was tossed off and hit the curbstone with his head. The doctor pronounced it a compound fracture of the skull, and refused to operate. Hal, characteristically, would not have it so and, stirred by him, we secured by telephone and general pulling of wires the head surgeon of the Denver & Rio Grande, who was given a special engine from headquarters. He performed the operation and saved Bill's life, and after a long convalescence, Bill went east. Unfortunately, thereafter he lacked something mentally. He lost that whimsical wit that endeared him to his friends and became rather a bore. He probably realized this and was, accordingly, depressed. Eventually he was given a job in Central America, and while there became obsessed with the idea that some one was trying to kill him. He was shot under mysterious circumstances, a sad end indeed.

Fourth of July brought the usual festivity. Throngs of people from the mines and outside tramped along the plank sidewalks. Cowboys in from the range rode the streets or tied their ponies to the rails outside the saloons. It happened that I, one Stevie, and Bob Meldrum, still a hired bravo, were walking along, gaily lighting firecrackers and tossing them under horses' feet to see them jump,— all taken in good fun. Meldrum, who, rarely for him, had taken quite a few drinks, held a cracker against the neck of a passer-by. It went off; the victim remonstrated heatedly. "Why, you son of a bitch," said Meldrum, "I hurt me own hand, what are you kicking about." With that, the other said, "No one calls me that name," and slugged Meldrum, flooring him. Meldrum was struggling with his gun, fortunately entangled with a Navajo blanket he was carrying, and friends who well knew the danger hustled the fellow away, still defiant and willing to continue the struggle. Stevie and I partially pacified Meldrum, but, to his mind, he "lost face," which his reputation as a killer could not stand, and he vowed to get his man sooner or later. Furthermore, he never entirely forgave us for causing his loss of prestige. The young fellow was a decent, valued employee of Tomboy Mine, and the vengeance simply could not be allowed. Meldrum had already killed a Tomboy man on flimsy grounds and was getting to be a prickly problem. The mine authorities took action and transferred him elsewhere, much to everyone's relief, including mine, who fancied a vengeful gleam in Meldrum's eye. It seems that these killers must be "blooded" every so often.

Our lease was still producing, sometimes at a profit, more often not, but it was apparent the end was approaching. Hal and I filled out our incomes to some extent by examination work but found it not very lucrative. Our trips were not without interest, however. One such was to Summitville, an abandoned camp over near the Rio Grande headwaters. The trip by rail being too circuitous, we took Brownie, our faithful pack horse, and astride our own chargers, Jamaica and Keno, crossed the Telluride divide at 13,000 feet, made the headwaters of the Rio Grande via Silverton over Wiminuche Pass into the Pine River valley to Pagosa Springs on the San Juan. This was all new country to me, wild and beautiful scenery, country which lingered in memory and through which, many years later, I was to lead family and friends, to find it still beautiful and unspoiled.

We finally made Summitville, examined the mine, which greatly belied the impressive assays we had seen, and set forth across country in the general direction of the Rio Grande. We had no trail, simply followed the gulches, only to find ourselves at dusk on a knife edge high up between two canyons. We camped in a place so narrow that we had to tie the horses head and tail to prevent their rolling off. In the morning we found a game trail which, eventually, led us to better travelling, but not before Brownie had taken a tumble which resulted in two or three somersaults, pack and all, to the bottom of the gulch. No damage except a broken fishing rod.

One examination was to Abiquiv, New Mexico, on the Rio Chama, an affluent of the Rio Grande. Here we were housed in Sr. Chavez, the *jefe's* dobe castle, into whose patio we rode and were made welcome—Mexican style. I renewed my Spanish with the Señora and her comely daughter, only to be taken aback, after a somewhat labored Spanish conversation, by the latter's remark, "How are the Yankees doing this year?" I remember seeing the glowing sheaf of Halley's comet in the southern sky, and the beauty of the river valley; no gold, however.

I had several other trips that year to mines in Colorado, Nevada and California, from which, needless to say, I was always glad to return to my wife and son in Telluride.

Our lease dragged on, but the end was in sight. Like all mines, danger was ever present. We expected one or two deaths from accident each year. The miners at this time were largely Greek or from the Balkans. Their names were beyond us and on the payroll were much simplified. One man was on our payroll as Bill Pants and answered to that name. His real name was Paputsis. About once a year some unfortunate met his death, usually underground. A favorite way to get killed was to overdump a loaded car into an orepass, and by the instinct to hang on when he felt the car going, the victim followed it down. Of course, drilling into missed shots was a rather frequent cause with dire results. The most perilous job was blasting mill holes. The ore from the stopes was shovelled down built-up rock mills and sometimes would hang up above the pocket. One man was deputed to the job of blasting the overhanging rock. To do this he often had to climb many feet up the narrow mill hole, light his fuse and make his getaway. Needless to say, he was paid double time, and the job was greatly coveted. Strangely enough, no one was killed at this work. Our only open-air accident was a curious one. The waste dump at the Sheridan had lengthened so that it was dangerously close to the high-tension wire. A man was put to picking down the dump, and was cautioned not to touch the wire. When he was relieved, he told his successor, "Don't touch that wire; it will kill you;" and so saying, touched it with his crowbar—I suppose for illustration. Of course, several thousand volts finished his career. The snow was deep, and a horse-drawn toboggan was used to carry him down. Halstead rode ahead to point the trail. Every once in a while the corpse would slide off the toboggan and roll down the steep hillside. Hal swore that with every roll, the corpse, which had stiffened with one arm extended, would point an accusing finger at him.

We leasers had one rule: whenever a mortality occurred, the partner in charge officiated as the firm's representative. A holiday was given, a procession of carriages was formed, and, headed by the hearse, next the priest accompanied by the partner, and finally, the long procession of mourners, genuine and pseudo, proceeded to the cemetery. After the burial, some companion wrote home proudly of the honor done the deceased, and that was the last we heard of it.

Toward the end of the summer our son and heir was not doing very well. He caught the whooping cough and, though he weathered that, still had a temperature and was losing weight. We became alarmed finally, and, the regular doctor being out of town, summoned the only one left in town, a woman doctor by the name of Brown. She diagnosed the case as due to the continued temperature, causing debilitation, and prescribed an ice bath to bring it down. Immersed in the icy cold water, poor Bobby's cries gradually quieted until, while taking a temperature, the doctor cried, "Take him out quick; he is near collapse." To do the fool doctor justice, her next action was prompt, and probably saved the boy's life. She wrapped him in blankets and popped him into the nearby oven until his temperature was normal; and Gwen and I breathed again.

Hal and I, seeing the end of Telluride activities, determined on a step we had long considered, to set up an office in Boston for the examination of mines. In the fall I sent Gwen and Bobby to Colorado Springs, and Hal and I wound things up in Telluride. We gave up the lease, our hopes of a new mine in the Ansborough, and abandoned hopes of developing "Treetops" as a going ranch.

In November, rather mournfully, I pulled out for Boston and the new adventure, leaving Gwen and the baby to follow. We settled in Dedham, the new office was established at 60 State Street, and the Telluride chapter was closed.

XIV

Canada

I had no intention of continuing my journal beyond the finish of Telluride days, but by force of circumstances I find myself consigned to a sedentary life as contrasted with the active one I had six years ago, and, wanting something to occupy my mind, turn to writing.

In August, 1951, I had a stroke (cerebral thrombosis), brought on by too strenuous work in a hot sun, which put me in the hospital and bed at home for a couple of months, and though I get around with the help of a cane, go to the office two or three times a week, and enjoy life socially to a limited extent, I have to consider that mining trips, shooting, skiing, chopping trees, etc., are over.

Hal's and my attempt to set up a Boston office was not very successful. He came down with an acute attack of rheumatic fever which put him out of the running, so that he never was able to join me in Boston.

I rented a small office at 60 State Street and carried on alone, but found clients not very eager.

Gwen and I rented a small house in Dedham, in which we were installed for the time being. My first professional engagement came from my friend Bill Nickerson, who sent me in behalf of his firm to examine the Kerr Lake Mine in Cobalt, Ontario. This was then what was considered on the edge of the northern wilderness. We little knew what a wealth of mines would be found far north of this.

Cobalt was a populous place, just then buried in the snowy heart of a northern winter. Shafts and headframes in the midst of dwellings and ice-covered lakes were everywhere. I found a receptive staff at the mine, perhaps extra jolly on account of the New Year festivities still in progress.

The mine was very rich, narrow veins many of which were nearly solid silver, and I was able to give it a very favorable report. The district was not very old. It was discovered in the course of extending a railroad some six years before. The railroad was not primarily to open mining country, but to reach the so-called clay belt a few miles north, for the benefit of farm settlers.

The veins in general were not very wide and went only to a moderate depth, but for a few years, while it was at its peak, the camp was one of the richest silver districts on the continent. Often carloads of ore sent to the smelter would assay four thousand ounces to the ton.

Again my friend Nickerson engaged me for an examination of the Planet Mine in Arizona, one of the Lewisohn's discards in their search for disseminated copper ore.[1] I misunderstood the problem, but fortunately didn't report favorably enough to lead my principals astray.

My first year in Boston as a consulting engineer was not a prosperous one. Apparently the demand for young, unknown engineers from the West was not overwhelming. I spent hours cooling my heels in New York offices, and landed a few jobs, some unpaid and none lucrative.

I remember one trip (expenses only paid) to a desert property not far from Goldfield [Nevada], which, incidently, was a flop. My chief recollections of that trip were the forlorn aspect of the formerly flashy camp with some fading splendor in a several-story hotel at which we stayed which still supported an elevator with a uniformed operator.

The trip back from the mine through the desert was quite memorable. The car broke down at intervals and we would build a sagebrush fire for light. Looking back over the long slope, we could see a succession of fires denoting each stop for repairs, perhaps ten in number.

I stopped off in Telluride and sadly wound things up as far as

---

[1] The Lewisohn brothers established an import-export firm in New York in 1872. Over the years they developed extensive mining interests in North and South America, in 1898, with William Rockefeller and H. H. Rogers, founding the United Metals Company. Adolph Lewisohn was well known for his various philanthropic activities including the building of the famous Lewisohn Stadium of the City College of New York.

possible, leaving the house and ranch to fate. (The house was sold for practically nothing, but the ranch was bought by Lavender at a handsome figure.)

One of my poorly paid jobs was a trip to Sonora on behalf of one Steindler, whose son, a New York stripling whose chief interest throughout the journey was the baseball fortunes of the Yankees, accompanied me for educational purposes.

Steindler had two properties in Mexico, both bad bets. He probably wanted to find out just how bad they were and availed himself of my cheap services. We took a stage from Ysabel,[2] a station on the rail to the Tigre, then a prosperous gold mine, where we were well received by Budrow, the manager, and lent mules for the further trip. The only incident of note was the encounter of a band of insurrectors, armed and fierce of aspect, their leader girded with a sword. But their approach was mild, resulting in a gift of cigarettes only.

Temblor, the mine, was insignificant, though rich enough in spots to be a "teaser." Thence we went by rail and wagon to Antgina, another moribund mine. This was in the midst of an arid cactus desert. We were sitting exhausted by a rare spring, when I spied in the tree above us a plentiful crop of limes. A bottle of gin was broken out of our pack, and a refreshing gin rickey was the result and our reviver.

At the mine, which had once been quite active, we took refuge by evening, after hot and tiring work underground, in the tank, which still held two or three feet of water and afforded us a refreshing swim.

I delivered young Steindler to his father, collected my small fee and put a period to the trip.

Hal, who by this time had recovered, and I fooled around with various mining moguls in New York, who had little to offer except intimations. I looked over my old friend the Camp Bird with a view to leasing and found it as cleanly worked out as could be seen or guessed. Scott and Beaton received me as an old friend, but sadly acknowledged that there was nothing left. Concealed from us was that remote and apparently barren part of the mine which years later developed into the prosperous King lease. Old mines truly die hard!

[2] Ysabel, no longer in existence, was about eighty miles southeast of Nogales, Arizona.

Hal, tiring of holding down office chairs, finally got a job as manager of the Wettlaufer, a nice little mine a few miles south of the Cobalt District, and took up residence there.

Shortly after, in one of my New York quests, I happened to pay a call on Sam Rosenstamm, one of the Lewisohn staff, and found him discontent with the Kerr Lake management. "How would you like a good man to replace your present manager?" said I. "Who?" said Rosenstamm. "Me," said I, and the upshot of it was that I got the job.

In January, 1912, I left Gwen and the babies (by this time Cecily, our second, had arrived) in the McMillan house in Dedham and went to Cobalt. I found that the Lewisohns had not established me officially as manager, simply had notified Seward that he was fired. As a consequence, I found myself a pseudo-guest in the house along with the Sewards, who had no apparent intention of leaving, and their four children. Seward did not take his congee gracefully, but continued to strive for re-establishment by long absences in New York, leaving me in a somewhat equivocal position.

I found that there was little need for concern as far as the mine went. Seward had neglected stope development, and, finding ore rather inaccessible, had tried to play safe by minimizing the ore reserves.

My chief job was to restore New York officers' confidence in the mine. Other preliminaries were to get acquainted with the staff, scout around the district and to report on other prospects in the neighborhood, in which the Lewisohns had an interest. One such, a trip by winter stage and snowshoe to Gowganda, an outlying silver camp with rich but erratic showings, interesting but unproductive.

Meanwhile, the community held off and watched me with a speculative eye to see if I would prove acceptable.

In early April, I was joined by my family, Gwen and the two blooming children, and domestic life began again.

By this time I was fairly well known by the social community, and accepted, I hope, and Gwen's arrival clinched matters. From now on, for nearly five years, we were settled in our new mining-town life. We found several worth-while friends whose pleasant acquaintance lasted many years. Cobalt attracted interesting people from over a good part of the mining world, Canada, U.S., and New Zealand, England, etc. The mines were prosperous and social life active.

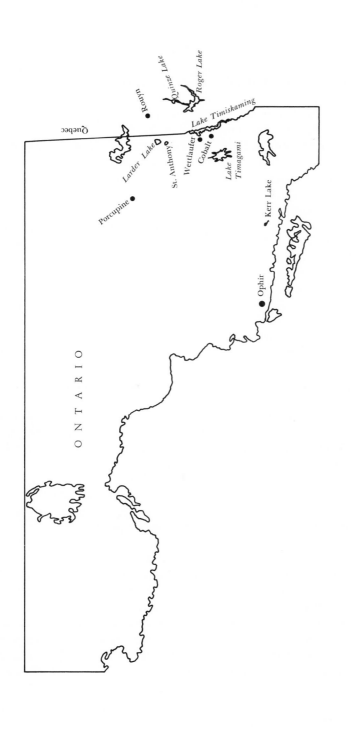

Rouyn

Quinze Lake

Roger Lake

Québec

Lake Timiskaming

Larder Lake

St. Anthony

Wettlaufer

Cobalt

Lake
Timagami

Kerr Lake

Porcupine

ONTARIO

Ophir

Toronto, four hundred miles south, provided a source of supply for most luxuries, and the surrounding wilderness gave the outdoor-minded plenty of sport.

The mine, under the efficient captaincy of Paddy Fleming, of recent Irish antecedents, and with a fair share of that nation's wit and wiliness, went smoothly. I gave Paddy a free hand (which he did not have under Seward) to make the stopes accessible by square-setting, and thenceforth the monthly production was easily attained. A good deal of the New York office's uncertainty was caused by Seward's timidity by prophesying a short life to the mine. I made a careful estimate of the ore reserves and restored some of their confidence. By the summer of 1912, it seemed as if everything would go smoothly, but the situation was changed for the worse in that Julius Lewisohn, son of Adolph, was made president and almost immediately made himself felt. He was a most arbitrary person.

To add to the annoyance, who should turn up but Seward, who I thought had vanished from the scene. He came at Julius' instigation, ostensibly to check on the ore reserves, stayed around ineffectually and departed after a few days, then after an interval appeared again! It seems that Julius was disgruntled at not having been consulted and was working to reinstate him and fire me. To add to the joy, telegrams arrived at intervals such as "When is Livermore leaving," etc., but, backed by a loyal staff, I continued my superintendency, and Seward eventually departed, to appeal to New York, no doubt.

To add to my tribulations, Lewisohn returned, made himself generally disagreeable, and at last, when given absurd and arbitrary orders in the mine, I blew up, offered my resignation, and went on surface to nurse my wrath. To my surprise, Lewisohn became almost affable; and only then it dawned on me that my resignation was what he had been gunning for right along. I determined to go to New York and lay my resignation before the board of directors, and with the aid of a loyal staff, boarded the carriage minutes before the train left, leaving Lewisohn surprised and baffled behind. He almost pleaded with me not to go, but to no avail, and as I departed, I had the satisfaction of telling him just what I thought of him and his methods.

In New York, I put my case before the board and offered my resignation again, but was prevailed upon to stay on till the annual meeting in September, to which I agreed on condition that Julius should have no more to do with me or the mine. It must have been a hard dose for the old man to swallow, but he did with dignity, and I returned to Cobalt.

As a windup after the meeting, I got telegrams of congratulation confirming my position and announcing the election of my friend Bill Nickerson, instead of Lewisohn, who resigned—the end of the battle.

During the summer, aside from getting acquainted socially and otherwise with the district, Gwen and I took a trip by steamer down Lake Timiskaming to visit Hal and Lorna at the Wettlaufer Mine, a rich little property about seventeen miles south. Some time since, while Hal was away, their residence was burnt up and Lorna had a pretty strenuous time. At the time of our visit, a new house had been built, and we had a cozy time, cocktails, a swim, and talking over old Telluride days. Shortly after, Hal got a job in Colorado and I was delegated to add Wettlaufer to my string.

The country north of Cobalt had been opening up very fast. The woods were full of prospectors. The gold camp of Porcupine had been established several years. At least two mines of major importance, Dome and Hollinger, were in full swing. McIntyre, afterwards one of the greatest, was temporarily in the doldrums.

I like others had inklings that more mines were to be found in the vast, almost unexplored northern Canada and urged my principals to form some kind of an organization to prospect. I did as much as I could by foot, snowshoe and canoe to visit the new discoveries, but could not do much alone.

The north country was opening up pretty fast. The T. & N. O. R.R. and the comparatively new Transcontinental, afterwards the Canadian National, had made things much more accessible to the prospector. Had it not been for the war of 1914, years of delay would have been avoided. As it was, several districts had been located, among them Kirkland Lake and Rouyn in Quebec. The former I visited, and camped in a tent with Harry Oakes and his pals at the Tough Oakes Mine.[3] Lake Shore was not yet tapped, and from it

[3] Stubborn and irascible in temperament, barrel-chested and short in physique, Harry Oakes (1874–1943) had a fantastic career before his untimely death in 1943. He was born in Sangerville, Maine; attended Bowdoin College; and then joined the Klondike Gold rush. The next decade he vagabonded around the

Oakes made his fortune. At Rouyn, where Noranda was afterwards developed, a great mass of barren sulphides defied development. I finally took an option on the Smith Labine claim and built a camp and sunk a shaft, but to no avail. All I got out of it were some trips on skis (then a novelty here) from the rail through the woods to the mine. The Lewisohns were still obdurate about forming an organization. The McIntyre, then barely started, was idle, and an option was being peddled around at $100,000, with no takers. I among others was offered the option and hopefully wired New York to take it. The answer was, however, "Is there enough ore in sight to pay for it?" Of course there wasn't, and my recommendation was turned down.

I tramped to the Larder Lake prospect and found it moribund. A surface showing of magnetite and other minerals, a speck or two of gold, but far too low grade to consider, at least in those days. This was afterwards the Kerr Addison Mine, one of the biggest and richest in Canada!

In one of my trips, I remember that the road by which I came in had been crossed by a rather lively forest fire, and I had to make a detour through tangled brush, wondering if I could make it before the fire swept up to me.

This northern Canadian country was a facer for the mining engineers used to the visible outcrops more usual in the States. The surface showings were often insignificant, the country covered with thick forest or muskeg, and the overburden heavy. Outcrops were so rare that even barren country rock was staked as a find. Henry

---

world, always in search of the "Seven Cities of Cibola." At the age of thirty-seven he discovered the Lake Shore mine in northern Ontario. Within a few years this bonanza became the main basis for his $200,000,000 fortune. Again Harry Oakes toured the world, this time in style, purchasing estates in Montreal, Bar Harbor, Sussex, Palm Beach, and finally Nassau. Attracted by the low tax rates and the pleasant climate, Harry Oakes favored the Bahamas for his residence. On the evening of July 7, 1943, after a fast game of Chinese checkers (which he won) Oakes retired. The next morning he was found in his bed, murdered. The suspects were as many as his numerous enemies. Today the murder is still unsolved, and its political repercussions reverberate through the Bahamas (John Kobler, "The Myths about the Oakes Murder," *Saturday Evening Post*, CCXXXII [October 24, 1959] pp. 18–21, 82–91; *Newsweek*, "Death in the Bahamas," CXII [July 19, 1943], p. 31).

Krumb, a visiting engineer, said sarcastically, "Gold is worth $50,000 a speck." Nevertheless, the geology became better understood, finds were made, and the Pre-Cambrian eventually became the Mother Lode of Canada.

At the end of the year, Paddy Fleming left for a more lucrative business, and I called on my old Camp Bird foreman, Bill Beaton, who was at loose ends. He arrived and afterwards sent for his family, and soon accustomed himself to the mine and the country. That hardy Scot was with me until he died. He had lost an eye years before, but the miners used to say he could see more with that one left than most of them would with two.

Gwen had stayed in Boston while the children had their adenoids out, and for general relief from a winter in Cobalt. I found time to improve my acquaintance with people and surroundings. Among other episodes, several of us, the Finucanes, Charley O'Connell, with whom I had become very friendly, visited Kirkland Lake again, to find it had grown amazingly, as promising mining camps have a way of doing. Although the Tough Oakes, one of the first discoveries, was still the best one. The Lake Shore Golconda was still sleeping.

As spring advanced a lot of us indulged in sport, as distinguished from mining, and went to Lake Timagami, a few miles south, even then a well-known fishing resort. We had wonderful fishing, with various varieties caught, bass, walleyed pike and lake trout.

The summer of 1913 was calm mentally and pleasant from a weather standpoint. Tennis was a favorite pastime, picnics on neighboring lakes and swimming in Lake Timiskaming, which at times assumed the air of a miniature bathing resort. The first motors arrived and carried on at what seemed terrifying speed over the limited good roads.

Forest fires engendered by the dry summer devastated a lot of the country north of us and even threatened outlying districts of Cobalt.

Our next-door neighboring mine the Crown Reserve lay almost entirely under water, and we, who lay partially submerged, and they had urged draining of the lake. After much hesitation, our superiors consented to the rather novel undertaking, and that summer a barge was constructed, on which were mounted efficient centrifugal pumps, a twenty-inch pipeline connected to carry water to Giroux Lake, and by September the pumping was well under way. At last the menace of flooding the mines was removed. This was a real danger. Mining the very rich ore too near the lake bottom might at any time break through and let in the flood.

I recall a rather ludicrous episode during Lewisohn's regime. He entered a drift extending out toward the lake bottom, which had been stopped at the danger point. A rich vein was still in place in the face. "Why don't you mine this?" said Lewisohn. "Too near water for safety," I replied. Lewisohn still insisted, when Paddy, with a wink at me, tapped a wooden plug left in a drill hole which had penetrated to water. A stream with the pressure of one hundred feet of lake burst forth, and all of us rushed to a safe spot. "What was that?" gibbered Lewisohn. "The lake," said Paddy. I don't know whether or not New York was influenced by Lewisohn, but for once he was on my side.

As mentioned, the summer of 1913 was a very pleasant one. We had made many friends, the Watsons, Finucanes, Fishers, O'Connells, Parkes, and others, who interchanged social events. Dinners, with Toronto's choice vintages, usually dancing after dinner—turkey trots were then the rage—a gramophone for music, sometimes a bit of gala with costumes en tête. Gladys Young visited us and enlivened the gaieties for a space. Then, all too soon the chill of autumn fell.

Meanwhile, the New York office had taken an option on the Hollinger Reserve Mine in Porcupine, and directed me to examine and tell them whether to exercise the option. The mine was fairly well developed, but had never made production. I was much engrossed with draining the lake, but got together a team of engineers and laid out a thorough system of sampling. The samples were assayed at Kerr Lake, and as the returns came in, they were consistently high and uniform—undoubtedly pay ore. New York urged acceptance of the option, fifty thousand dollars, but I was suspicious; the ore didn't look that good. I told New York to hold off while I resampled the mine. This time I did as much of the sampling myself as possible. The returns were nearly worthless. I could hardly believe it, and announced that I would make a mill run. This I did in a rented mill and found, as confirmation of my suspicion, that the mine was valueless. There was nothing to do but wire New York that I had been "salted," and advise dropping the option. We put detectives at work to find out how the deception was done, but to no avail, except that the expected chief beneficiary shortly left for other parts. I felt humiliated by this, but New York took it in good part and I lost none of their esteem. After all, I found out the fraud.

Draining the lake proceeded, and by fall most of the water, some 400,000,000 gallons, had been pumped out. One incident I remember, when we turned on the pumps the first time and a lusty stream issued forth, we were all elated, then shut down for adjustments. The pipeline, a twenty-inch spiral riveted one, led from the barge over a height of land. On shutting down, a vacuum was caused, and several lengths of pipe collapsed as flat as a pancake. This operation, being a novel one, was watched by many skeptics, and I just couldn't admit failure, so I summoned my mechanical staff and replaced the collapsed lengths in the dark. By putting in a vent standpipe, we corrected this error and all was well.[4]

Every morning, as the water lowered, I would make the round of shore in my canoe, a fascinating pursuit as the veins uncovered. Sometimes a new one would show up. One such, the top of a rather barren vein underground, was really spectacular—some thirteen inches wide and about half solid silver. This was afterwards known as the silver curb, and a section was quarried out and sold (not given) to the National Museum. Another preserved under the lake waters was polished and grooved by the glacier in the silver.

As winter advanced, the hunting instinct overcame me and resulted in Gwen and I taking our little cedar canoe and embarking on the Timiskaming steamer for the Montreal River, where we were put ashore to search for moose in the hinterland. The dam near the mouth had flooded what was originally a string of small lakes into one big one with banks under water and trees half drowned. We camped a couple of nights where we could find rare dry spots, and after encountering a half-gale of wind, with no refuge on shore, gave up and gameless returned to Cobalt.

A trip to the States and civilization was indicated after that rough outing. We left the children with the incomparable nurse, Elise Brouard, who had responded to our advertisement in the Montreal paper. Elise was a real treasure, a native of Guernsey, several cuts above the servant class. She was with us for many years, and still corresponds with the whole family. She went back to Guernsey after

4 The engineering trials and tribulations of his Cobalt experience are recounted in Robert Livermore, "Mining Districts of Northern Ontario," *Mining and Scientific Press*, III (January 15, 1916), pp. 89–92.

the children were grown, and endured the privations of the last war, alleviated a little by food and clothing which we were able to send. With Ella, our half-Indian cook, we felt well staffed.

Anticipating a long winter siege, we did up Boston and New York rather thoroughly, taking in the Yale game (Harvard won, for a wonder), the Army-Navy game in New York, a trip to Philadelphia to see our friends, the Hopkins, a ride or two around Westwood, and a memorable jumping skylark with the Rose Tree hunt. I found the old knee grip was still functioning on the robust hunter lent me.

In New York, we saw a few plays and night clubs, saw the Castles do their famous walk, and in Boston saw our friends and family, took in a dance or two, then back to Cobalt.

Shortly after our return, I took a trip to a new discovery in McArthur township, a hard trip, and an uncomfortable night in a single-room log cabin, filled with snoring, smelly prospectors. The contrast struck me forcibly; only three days away from top hats and dress clothes to a snowy waste in the north country, trudging over frozen lakes on snowshoes, towing a toboggan with equipment!

Before the freeze-up, the lake was almost drained of water, leaving a bottom of semiliquid mud, in places forty feet deep. The problem of getting rid of this was met by hydraulic nozzles near the pump intakes, which liquified the mud to the extent that the pumps could handle it. That job had to be postponed till the next spring season.

As the lake drained, thousands of fish were suffocated, and the effluvia from their carcasses bade fair to become a pest. One day, however, a lone gull appeared and hovered over the lake and took in the feast. He vanished and next day reappeared with several followers. By the week's end, a throng of gulls were present, and the fish nuisance was soon a thing of the past.

After the lake was completely drained, upon editorial urging, I wrote up a description, which appeared in various technical journals. My only claim to being an author.

Winter descended upon us as usual, but in spite of it we lived comfortably, with occasional social gatherings to enliven the ennui. Inspired by somewhat warmer weather, I wired Grace to visit us in late February. My telegram to her read, "The worst of the cold weather is over." Grace, with Ruth Nickerson, arrived and were met at the station by sleigh and comfortably ensconced at the house. Next morning, I suspected that it was rather cold and looked at the thermometer. It read 50° below! That telegram is a family cliché to this day.

I had taken up skiing again, inspired by the steep if short slopes made by the drained lake bottom, and found it good exercise in the long, snowy winter. When the Wettlaufer shut down for lack of ore, I imported most of the staff to Cobalt and found work for them. Among them was the bookkeeper, one Con Thomson, a Minnesota Norwegian, who had been an expert jumper. He got the company carpenter to build a take-off tower on the brim, and a jump on the steepest slope. I took one look at the layout and said, "Con, you go ahead; I'll never negotiate that." Con laughed and said, "Oh, yes, you will; just watch"; with that he took off and sailed through the air seventy feet or more.

I couldn't afford to lose face, and with inward trepidation climbed the tower and launched myself into space. To my surprise, I landed rather easily and covered thirty or forty feet. After that, the sport became a major distraction.

About this time, I had my first encounter with the gout. It occurred at intervals during the succeeding years, and has been an embarrassing and painful handicap all through my active life. So rare was it that neither I nor the resident doctor recognized it for what it was, until by a process of elimination we pinned it down.

With the coming of spring, though winter was still upon the land, we decided to emigrate with the whole family to the States for a change, but found the Boston climate hardly better than Cobalt's. After a few medical and dental attentions, a theatre or two, and a nonconclusive call on Mr. Hammond (with an eye to the future), I left the family at the rather depressing apartment on Commonwealth Avenue and returned to Cobalt.

I should add that the Wettlaufer Mine, which, as mentioned, fell to my management after Hal departed, necessitated a trip a month to oversee. In summer it was a pleasant sail by steamer, seventeen miles down Lake Timiskaming, but in winter it was quite an expedition. Ned Fournier, our coachman, brought out the low French-Canadian sleigh, and, bundled in furs, we drove over the ice, pleasant enough in fair weather, but I recall one trip made over the rough snow on shore, with the wind blowing a gale and the temperature 17° below. When I got home, though not frostbitten, my face was so stiffened that Gwen hardly recognized me.

Before facing the rigors of Ontario, Gwen and the children joined Mrs. Young and Gladys, and went to Warrenton, Virginia, for a taste of southern spring. Gwen said that their landlady, a Mrs. Britton, was an ardent Rebel sympathizer, and when she told her that her father-in-law had campaigned around Warrenton, her attitude became rather glacial. When Gwen afterward told her then that her sister-in-law was the granddaughter of Jefferson Davis, the temperamental climate was decidedly bettered.

The last of May, the family returned, and Cobalt's "summer season" started. At first, life was especially gay with dance dinners, picnics, swimming, tennis, etc.

Father visited us, as he always did to his sons, no matter what far places they inhabited and at whatever inconvenience to himself. When he left, I asked Beaton, who had accompanied us underground, "What did you think of my father, Bill?" Fixing me with his one blue eye, he said, "Ye'll never be the man that he is." I took this as a genuine compliment.

In August of that summer, an obscure royalty in central Europe was assassinated and the world exploded. Truth to tell, we weren't much excited, only rather pleasantly interested, at least at first. Of course, England would soon put the quietus on the "cowardly" Germans and things would go on as usual. However, we were quickly affected. For the first catastrophe, there was no market for silver. To my dismay, I received a telegram from headquarters to shut the mine down. After a futile exchange of telegrams, I had the disagreeable duty of firing many faithful employees who now found themselves classed as "enemy aliens." I managed, at some risk to my job, to keep the important members to keep the organization intact and essential work going, and after a few weeks the confusion quieted down, the silver market was resumed and things went on as before.

In our remote corner of Ontario, the war seemed far away, and after the first flurry, the usual activities continued. In September, a hunting party was organized consisting of the Finucanes, the Henrotins and us. The destination was the Quinze and Roger Lakes in Quebec. The Finucanes were close friends, the Henrotins recent acquaintances, who, it proved, did not wear too well. At the last minute, Gwen and I were delayed by Bill Nickerson, the president's, serious illness, but as it became apparent that it would be a long and lingering one (it was fatal finally), we decided to go and chance finding the others.

We took a launch at Haileybury with canoe and equipment to Ville Marie at the head of the lake, and at the Timiskaming Indian settlement picked up our guide, whose anglicized name was Jim Stainger, and set out by wagon for Lake Quinze.

After a night at Clarke's farm at the foot of the lake, and after un-availing attempts to start the motor on the farm launch, we set out for the long paddle on the fortunately calm lake and made the mouth of Roger River, where an Indian told us, by Jim's interpretation, that our party had proceeded.

A lengthy paddle and several portages up the stream brought us to the great Roger Lake, whose forest-clad shores we searched for the camp and at last found it on a sandy point; a relief to reach our goal in this wilderness, and general rejoicing by all.

For several days, we loafed, ate, slept and canoed, hunting at intervals. A moose came out on the shore near the camp about dusk and Ray took a shot at him, but missed, being unable to set his sights. Gwen and I, with Jim, took several canoe expeditions to various bays on the lake, but aside from seeing a distant moose and enduring a very rough paddle across a broad bay, returned gameless, except for a few partridges. Tiring of desultory hunting, I took Jim and the other Indian guide, Joe McKenzie by name, and made a determined hunt, taking a little grub and a couple of blankets in case of delay. We went where Gwen and I had seen the moose, and established ourselves on a little island near shore. Night came on and we were about to pull up, when a crashing and splashing accom-panied by bawling sounded as if all moosedom had arrived. It was too dark to see anything, but we paddled as near the commotion as we thought safe and finally saw a vague shape, at which I fired, to be rocked by fleeing moose, but none left behind. We made a nest of sorts in the brush and spent a restless night and, finding no moose on shore, made our way back to camp.

Several days of camp life, but gameless, made Gwen and me a bit homesick for the babies, and the party, influenced somewhat by Henrotin's uncertain temper, embarked for home. At the little Roger Lake portage, Gwen and I decided to stay in the woods for a last try for a moose and left the others to make their way home.

We camped a couple of times on the Roger River and Quinze, finding Jim a genial companion whose culinary art, camp style, was not to be despised. For instance, his beans cooked in a lard pail overnight in the hot ashes were a masterpiece. He was intrigued by Gwen's custom of stewing part of a lemon in the dried prunes, which appealed to him so that he experimented with lemon in the coffee!

In spite of Jim's appealing calls on the birchbark horn, no moose appeared, so we set out for Clarke's farm and home. Joe, after having delivered his party, inspired either by more hunting or the congenial company, rejoined us. At Clarke's, we decided to separate, Gwen to go home to the children; and I, determined to get that moose, took the two Indians, one canoe and a light outfit, and by a series of small lakes and streams headed for Bass Lake, good moose country, according to the guides.

With light equipment, one canoe and three paddlers, we made good time, by small lakes, beaver-dammed streams and many portages, to our destination, Bass Lake by night. I asked Jim, the Indian, the name of one lake, I remember, and he replied, "Obobskidae," the lake with the second-growth birches. Bass Lake was a pretty, clear body of water, but alas, a group of Indians had hunted out the surrounding country pretty thoroughly and prospects were not good.

I liked these Indians. It was quite different than in the more formal camp where the guides did most of the work and had their separate existence. We ate together and sat around the fire together with no feeling of "sport" and guide. One night as we watched the moonlight through the trees, I said, "Jim if we had a few more days and nights like this, I would be half Indian myself." Jim replied, "You are all Indian already," which I took as a compliment of high order.

A council of war developed that there was a small lake two miles distant where it was thought that there would be a good chance, so, cutting our scant equipage still farther, we set out through the trailless brush, one carrying the canoe, another the bedding, and me the rifle and what was left. We reached the lake, a lonely spot, and after a circle of the shore, finding fresh tracks, made camp and prepared to wait.

Night came on with no response to our calling and, discouraged, we made our fireless camp in the brush. I went to sleep, but about dawn the grunts of a bull moose were heard, and, half asleep, I got in the canoe, while the two Indians paddled quietly across the lake. In the dawn mist, the moose appeared looking wild and uncouth, staring a moment at us, then turning to flee. I fired and at the third shot down he came—at long last the end of the chase.

A fire on the beach, a breakfast, whose *pièce de résistance* was fresh moose liver, revived us after a chilly night in the bush, and with the head, hide and what meat we could carry, we made our way rejoicing to Bass Lake and later the journey home. The moose antlers were not big, but satisfactory, and they now adorn my front hall. I remember as we passed through the Indian village on return, a youth called out to Jim in the Indian tongue. "What did he say?" I asked. Jim translated rather sheepishly, "Where did you get him? Did you catch him in a trap?"

The remaining days of 1914 were rather an anticlimax after the moose hunt. The whole family went to Boston early in the winter, and as another arrival was expected by Gwen, we installed ourselves at a nice little apartment at 32 River Street, while I occupied the time by business trips to New York, meanwhile, with an eye open for future advancement—finally returning to Cobalt for a bachelor's winter. The routine of work, varied by an occasional trip to look at a bush prospect and one or more visits to the States on business, passed the time rather quickly. The war became increasingly personal as familiar faces disappeared for the Canadian Army. Then came the *Lusitania* sinking. At last it has come for us, we thought. The United States can hardly stay out now!

Shortly afterwards, we were called for a meeting in Cobalt to hear an important speech by President Wilson. I forget how it was transmitted, but as Wilson's polished periods were rounded, we listened eagerly and attentively. As the speech went on we wondered when he would express his indignation and even a call for war. At last came the climax. "A nation can be too proud to fight." One by one, we Americans sneaked home decidedly disillusioned and feeling a bit like a cur retreating with his tail between his legs.

As spring came on, relief from winter's snow and cold made amusement, even trivial, a necessity. There were several wifeless husbands like me, looking for distraction. Bridge was one, a punching bag at which I took daily exercise another. Someone started kite flying of boyhood memories. All of the Kerr Lake staff made the conventional models, with which we vied for size and height at the ever windy lake basin. We found the limit was reached by the weight of tail and string and were about to call the sport off, when the Chinese cook appeared one day with a flimsy butterfly-shaped affair, which he launched on á zephyr and controlled with a spool of thread which soared far above our logy contraptions. We retired in confusion to be beaten by China.

In April, I went to Boston again, partly to visit 32 River Street and partly to scout for future connections, and again in June. As the new baby was expected shortly, I took in the fifteenth reunion of my class, an outing at Plymouth and about the last gasp of athletic activities, baseball, golf, etc. Seeing the grey hairs and balding heads of some of my classmates, I was unpleasantly reminded of the approach of old age.

On June 15th, our youngest daughter arrived with little turmoil, and henceforth became an equally important member of the family.

Leaving my augmented flock, I returned to Cobalt. During my absence a terrible accident had occurred at the mine, one of the few casualties that happened during my incumbency. My mine engineer, a fine young fellow named Scott Eldredge, with a visitor to whom he was showing the mine, traveled by an unused opening to the main workings and were met by the noon blast in full force. Both were instantly killed, of course. Naturally, it put a damper on our spirits for a long time. Investigation by the authorities decided that it was pure absence of mind on Eldredge's part and held us blameless. Looking back on it, however, I think we should have sealed off that long-unused opening. It is hard to guard against everything.

Gwen rejoined me in due course, and things went on apace. As a result of my talks with Jennings of the U.S. Smelting Company, they sent up one of their engineers, one H. S. Lee, to look over the north country under my auspices. Lee was a nice chap, formerly running a mine in Oregon, and together we made a trip through several of the other camps to the north and various newly discovered prospects. The trip was as much for my education as Lee's. I hoped for some connection with the U.S. Smelting as a result of my efforts, but that cagey company had no such intentions, as it developed later. Lee, who had sporting instincts (he was quite a well-known football player in the West and was known as "Tubby" Lee), and I made an excursion to the head of Lake Timiskaming on a duck shoot, at which we had some success due to the ingenuity of our guide, who made decoys out of bent tree roots blackened in the fire. He reported favorably on the prospects to his principals, but the only result as far as I was concerned was that I was put off by polite evasions, and Lee was sent in as the U.S. representative. I found that I had been a patsy for the U.S. Smelting people.

In one of my trips through Toronto, I witnessed the march through of 18,000 Canadian infantry, the first trained unit to be sent to Europe. They were a fine-looking, hardy lot. Among them was the so-called "American Legion," a regiment distinguished by a small American flag on their tunics, in appearance otherwise no different than their Canadian comrades. We Americans took a bit of pride in the fact that some of us were willing to join the right side.

An amusing episode occurred during the summer. Some time before, in Cambridge, two young college friends, Jim White and Bill Claflin, announced their intention of traveling by canoe to Hudson Bay and the Albany River. One day I got a telegram from far up the line, "Arriving Cobalt by train, will you pay our fare." In due course, two bearded, moccasin-clad young men appeared on the Cobalt platform, their railroad fare liquidated, and taken home to Kerr Lake. They stayed with us several days, welcome guests, and ever since I have counted them my friends. Their tales of life with the Cree Indian were as good as a book.

During the summer, the Lewisohns took on two properties, again without asking my opinion. These were the Cobalt Comet, an adjoining property, which was nearly worked out, at a price of fifty thousand dollars. This mine was formerly the Drummond, named for the poet, and the subject of the French-Canadian dialect verses "De vein calcite." The other was a gold mine, called the St. Anthony, far to the north on Sturgeon Lake. It had been worked for some time with rather poor success, though rich in spots. I appointed a superintendent, Mark Little, and later took in a staff consisting of a bookkeeper, surveyor and blacksmith to join him. Access was by the Transcontinental R.R. to a tank station called Buckes, and then a hike through the woods to the lake, which was supposed to be frozen enough to bear. An early snow had covered the muskeg, so that swampy holes were concealed, and we were soon wet to the knees and half frozen. On reaching the lake, some eight miles, we found that it was open water, and rather than endure the slush, we waded along the shore, as an alternative to frozen toes. A choice of evils. We were glad indeed to reach the comfort of the warm mine buildings.

This was good moose country, and my hunting instinct soon blossomed in a trip by canoe in the surrounding waters guided by a willing miner, half hunter, Tabot by name. We camped overnight in a deserted log shack and scoured the woods around us, but to no avail, except to bag a fisher which we saw swimming in the lake. He made the mistake when he reached shore of stopping to show his teeth at us and fell to my gun.

This was on the edge of the Indian country, and evidence of their presence was to be seen here and there in the woods. I think one of the finest examples of an Indian habitation I have seen was a wigwam on the shore of the lake all made of squares of birchbark carefully fitted and cemented with pitch in the traditional V shape. I think the tribe was Ojibway. It was untenanted, as the family was elsewhere. One bad effect of the war was that there was little market for furs. The Hudson's Bay Company had apparently abandoned the Indians, who had lost their source of income and were living from hand to mouth. I came across a fisher in a trap very much alive and full of fight. Tabot said, "Take him and leave four dollars in the trap! That will be ample for the trapper." (Fisher pelts were ordinarily worth twenty or thirty dollars in the market.) But I contented myself by putting the animal out of his misery and left him for the owner.

That fall, I made an expedition down Giroux Lake with Ned Fournier, who beside being an efficient coachman, liked hunting as much as I and was a good companion besides. I finally got my deer, by a good still hunt in the light snow, which Ned and I dragged in over the ice right to the door of my house.

One other trip several of us made to Paradis Bay, on the shore of Timiskaming, where deer were said to be plentiful. We went in the Ford, as the ground was bare and the going good. Several deer were seen, but at dusk, and our aim was poor. Alas, the weather turned soft, the frozen clay of the road melted and it seemed as if we were marooned. Just then the *Meteor* was sighted on her way up the lake, and we hastily gathered up our equipment and made a dash for the landing, hoping to ship ourselves and the car. Unfortunately, the boat didn't see fit to stop and we found ourselves again marooned.

We were bemoaning our luck when one of our party, ever cheerful, produced a kettle of soup which he had retrieved when we abandoned camp, with the remark, "Have some soup." That restored our equanimity, so we made camp and in the morning set out overland. It was hard going, even for a Ford, but, aided by the combined manpower of the party, we at last made Cobalt and a welcome meal and bed.

During the winter which soon arrived and piled up its lasting snow, I found some relaxation in skiing, jumping in the lake basin, under Con Thomson's expert tutelage. I got pretty fair at it, but never equaled Con's graceful leaps.

We instituted a meet for the youngsters at which prizes were given, and some of them became fairly good, but the sport never became as popular as I hoped. Too cold, I guess.

The mine kept up its steady production, once the flurry caused by the beginning of war subsided. The Comet seemed about to die, when as a last resort, I had a trench driven in the one piece of ground untapped, a small piece right in front of the office which had been trodden for years. To my joy, we uncovered a rich vein loaded with silver. This was the nine-days wonder of the camp, at which everyone came to stare and to remark how the former owner had overlooked a bet so near home. Visions of recovering all the price paid for the mine were ours for the time being, but, sad to say, the find was only a flash in the pan and soon petered out.

A terrible accident occurred in this mine. MacFarlane, the superintendent, was killed by picking into an unexploded drill hole. Soon after, the mine was closed, all the ore having been extracted. It was an unlucky mine from the start, but it was not a complete loss. I believe we eventually recovered nearly all the purchase price. Anyway, it was not my baby.

In February, as a break from the winter, the whole family emigrated to Boston. We stayed with father in his new house on Bay State Road, and had the usual round of winter activities. It was a great pleasure to see father installed under his own roof, instead of the apartment, also the opportunity of his getting to know his grandchildren. I made a trip to the head office in New York, but found no new hopes of advancement, and, leaving Gwen and the children, returned to my bachelor existence in Cobalt.

In March, I made a trip to the St. Anthony, this time taking my skis. The weather was fine and I thoroughly enjoyed skimming over the frozen surface of the lake. Evidence of the northern environment were fresh tracks of caribou, which seldom appear south of here. Also, a beautifully matched dog team owned by somebody near the mine.

As spring reluctantly approached, the family returned. The children grew rapidly in strength, and were a constant source of interest and sometimes of apprehension. They were prime favorites among the head men at the mine, who probably kept them from dangerous activities. One alarm, I remember, when the bookkeeper rushed up to the house saying, "Bobby is on the roof." We found him on a steep, narrow slope outside an upstairs window, from which he had crawled perilously, perched on the edge of a thirty-foot drop. I stationed the bookkeeper below to keep him distracted while I quietly reached out the window and grasped him. Needless to say, we breathed again.

On the whole, the children were a constant pleasure, but, of course, admonitions were needed occasionally. In fact, once when Cecy was ominously quiet, a call for her by name brought the response "Nussing," advance notice of a query which would have been "What are you doing?"

This winter, I set Bobby on skis. He wasn't very proficient, but at least he got the early feel of them. However, I pride myself that he had a start for the champ he afterwards became.

That spring, our dormant love of riding overcame us, and I bought a nice little horse for Gwen and requisitioned one of the driving pair for myself. On them, we toured the limited highways of the district and ventured out into the neighboring trails. We found the frequent corduroy rather trying on the horses' legs, and progress slow, but managed to have a bit of distraction. However, Cobalt was not superlative as a riding district.

In spite of the war, news of which was not too good, life went on at Cobalt much as usual. The summer was kind; tennis, picnics, and swimming in Timiskaming passed the time pleasantly. However, I had a feeling of restlessness, partly because of hovering war clouds and partly because I felt that I had reached as far as I could go professionally, as far as Cobalt was concerned. Silver production was beginning to decline. The mines had apparently passed their peak. I couldn't see myself stuck in a dying corner of the mining world. I guess if I had stuck it out a few years, I would have benefited like most of the rest of my Cobalt friends by the rapidly expanding North, but the lure of newer fields attracted me.

Hal, who had returned from a physically distressing trip to Guiana, and was consulting engineer for the Goodrich Lockart Company, was anxious to have me join him in that company; and after several consultations in New York, I decided to resign. With some trepidation, I announced the fact to the Lewisohns, who offered me an advance in salary, then asked me to stay on till the end of the fiscal year in September.

I engaged Harry Kee, a most competent man, from Rolla Watson's staff to take my place, and to my satisfaction was offered the consulting engineership at a five-thousand-dollar fee.

With the congenial companionship, a salary of seven thousand dollars and an interest in the properties which we might find, I felt the future was fairly well assured.

This summer was a terrible breeder of forest fires. The bush was as dry as tinder, and soon the fires burst out. Several settlements and isolated homes were consumed. The flames swept over cleared spaces as much as a mile wide, and the loss of life was said to be as high as 250 people. One fearful holocaust was when some 30 people took refuge in a railroad cut near New Liskeard and were roasted to a crisp. The trains brought in scores of refugees bereft of everything, and we ransacked our houses to provide them clothing and necessaries. Bush fires were always a curse of the North.

One reason for Hal's wishing me to join him, aside from friendship, was that he was commissioned for the Army and wanted to introduce me before he vacated. The last of August, we moved from Cobalt, leaving the children, however, as Boston was inflicted with a plague of polio.

I was to make a trip with Hal of the various properties which Goodrich Lockart had taken on, covering mines from Arizona to Oregon, afterwards returning to Cobalt for a final check-up.

Gwen meanwhile had leased a nice little house at 23 Charles River Square, and by the time I returned from my trip they were nicely installed.

And that ends my account of nearly five years in Cobalt.

Epilogue

In the winter of 1915–1916, Robert Livermore returned to New York to consider his business affairs and his future. "I see forty years staring me in the face, but it doesn't scare me half as much as it might, as I am as far as I can see about as young in appearance as I was ten years ago, and certainly much more able, physically and mentally, also have a lot more to live for."[1]

By the summer of 1916, his long mental anguish over his Canadian future was over; as soon as possible he would leave the wilderness for the States. "After a long cogitation Gwen and I decided that longer residence here [would] lead to stagnation, and that the few thousand we might save during the life of the mine were not worth the pains of exile in a wintry place, when opportunity was continually passing by."[2]

The economic adjustment of moving back to New England was more quickly solved than Livermore could have dared hope. At the end of July, he was in the plush offices of the mining investment firm of the Goodrich, Lockart Company at 61 Broadway, Manhattan, talking over his future with Dave Goodrich. The negotiations culminated in Livermore's being offered a "handsome" salary, plus permission to continue consulting for the Lewisohn interests. Besides the financial advantages, he had the pleasure of working with his lifelong friend Halstead Lindsley, who was also a member of the Goodrich, Lockart engineering staff.

[1] Robert Livermore, "Diary," p. 16, Livermore Collection, Western History Research Center, University of Wyoming.
[2] *Ibid.*, p. 20.

As his first assignment, he accompanied Lindsley to the Southwest to examine various holdings of Goodrich, Lockart, especially in Arkansas, and an Arizona mine near Jerome, Arizona. By the winter of 1916, Livermore was again becoming restless: "I am afraid, however, that the firm is not really in earnest about becoming a real mining firm, as they are branching out in a dozen different directions, which all take enormous sums of money. . . ."[3] The passing of another six months only served to confirm his judgment: "The great trouble with this company is that they never have a firm policy about any of their mines but left them drift."[4]

The advent of World War I increased Livermore's mental turbulence. His family had a long tradition of patriotism. Yet he also had heavy responsibilities, and the bank balance was not great enough to cushion against a prolonged absence of financial support. Finally in July, 1918, he enlisted as a captain in the Fifth Engineer training regiment. Stationed at Camp Lee and then Camp Humphrey, Virginia, he mused often that he marched "over the very ground that my father had trod in the seige of Petersburg [during the Civil War]."[5]

After the armistice, Livermore retained a tenuous connection with Goodrich, Lockart, although more and more he turned his attention to the shipping firm of William H. Randall and Company, in which he had become a partner before his enlistment. The thought of a new economic enterprise was stimulating. Now at least the promise of greater financial remuneration than he ever experienced seemed at hand. Also in the spring of 1918, he purchased Boxfield, a delightful New England rural estate, between the towns of Boxford and Topsfield. Here he would return time and time again, weary and discouraged, to relax and discover the pleasure of life anew; those fifty-odd acres of tillage and woodland became his haven.

Through 1919 and 1920 several mergers and realignments in his shipping interests continued to keep his business affairs in constant flux. The first reorganization was the combination of the firm of W. A. Harriman with William H. Randall and Company to form Livermore, Dearborn and Company. Then a brokerage company, Rojas, Randall and Company, with its primary field of economic

3 *Ibid.*, p. 36.
4 *Ibid.*, p. 40.
5 *Ibid.*, p. 59–60.

interest centering in South America, was formed. Livermore's experience in the brokerage and shipping business, while unforgettable, was one of the more painful periods in his business career. On September 6, 1920, he brought himself to note in his diary, "The tale of woe of Livermore, Rojas and Co., [to] have to tell is too burdensome to me to allow myself the patience to write it." In sum, "suffice it to say that bad credit and business conditions of this year combined with various errors and lack of experience on our part have put us frankly speaking, on the shore perilously close to the rocks." [6]

In the latter part of 1921, Livermore lived a miserable existence in Argentina, trying his best to straighten out the tangled affairs of the brokerage business. Negotiations with the South American businessmen were made all the more miserable by the throbbing of a broken jaw, which Livermore incurred in a shipboard accident on his way to Rio. By mid-1923, the brokerage and shipping interests were liquidated, and Livermore was again in search of a rainbow. Temporarily he found employment with his old friend Halstead Lindsley.

Bulkeley Wells left the Telluride area in 1923, and as the New Year of 1924 approached, Livermore found himself confronted with the distasteful and seemingly impossible task of saving the Smuggler-Union mine in Telluride. He entered in his diary for January 31, 1924: "I am off to Colorado day after tomorrow, and am going to try to liquidate the Smuggler without bankruptcy, but it is going to be a difficult job." [7]

The Smuggler-Union had remained the main investment of the Livermore family's money, talents, and energies. Discovered in 1876, the Smuggler-Union had a history that caused its owners to alternate between hope and despair, [8] as did many other mining ventures throughout the West. When the Livermore-Aggassiz-Shaw syndicate

[6] *Ibid.*, p. 83.

[7] *Ibid.*, p. 98.

[8] The history of the Smuggler-Union can be pieced together from: *Colorado Mining Director and Buyer's Guide* (Denver: Smith-Brooks, 1901); *Rocky Mountain News*, September 21, 1902; Denver *Times*, June 15, 1898; Telluride Daily *Journal*, December 30, 1907; *Mines and Mining in Colorado* (Denver, 1893); Charles N. Bell, "Mining Methods of the Telluride District," *Transactions* of the American Institute of Mining Engineers (February, 1924), pp. 1–15; and Wilson Rockwell, "Telluride Story," Telluride *Times*, February 5, 12, 19, and 26, 1965.

assumed control in 1898, the mine had more than thirty miles of tunnels and employed approximately three hundred men. The Boston investors immediately set about on a program of modernization. By 1900, nine hundred men were receiving Smuggler-Union pay checks. The mill at the tiny community of Pandora was revamped. By 1907, a banner year, the Pennsylvania tunnel was finished, exposing one of the largest bodies of ore in the Smuggler. Of even more significance was the completion that year of a new generating system, whereby water power from Blue Lake above Bridal Veil Fall was piped over twenty-thousand feet. With ample electricity and new ore, the future of the Smuggler-Union seemed assured. However, the erratic market, strikes, and fires continued to make the annual meetings in Boston an embarrassing chore for the Telluride management.

The atmosphere in Telluride was pretty much as Livermore had remembered it a decade before, when he first skied across the brilliantly white San Juans. "Telluride, in spite of modern changes such as motors, remains just about the same old wicked but amusing place. The town is pretty wide open in spite of prohibition and the 'row' flourishes. The chief trouble is with high-graders, who are getting away with a good deal of stuff, presumably from us mostly. We have made one arrest, and though we cannot hope for conviction, we do hope to get the stuff in time. The other night, we got wind of a car coming up for high grade supposed to be cached near the mill, and lay out with guns to receive them till 2 o'clock a.m., only to hear they came at 4 o'clock."[9]

Dividends were forthcoming, but they seldom matched anticipated rewards. To the group of stockholders who had developed the amazingly lucrative Calumet and Hecla mines, the Smuggler-Union was a sad affair. Few mines in Colorado received as lavish investment as the Smuggler-Union, but few were more exasperating to management and shareholders alike.

Livermore had no more than settled down when the price of silver plummeted with the expiration of the Pittman Silver Purchase Act. By that autumn he was confronted with the alternative of either making the Smuggler-Union profitable or closing the mine. After an extensive analysis of the financial situation, Livermore recommended a four-point revitalization program to the directors:

[9] Livermore, "Diary," p. 105.

continue operations, liquidate all standing debts as quickly as possible, undertake developmental exploration to locate new ore reserves and modernize the mill, and increase production to the point of profitability.[10]

The necessity of continuing operations was made more obvious when it was discovered that the outstanding debts of the company exceeded the estimated recoverable assets. Within a year, the old debts were liquidated. This dramatic change of fortune came about through the successful collection of the "accounts due" page of the Smuggler-Union ledger. Livermore laconically commented that "these accounts were naturally paid more willingly to a going concern than to a salvage operation."[11] All efforts were now turned to the development of ore in the far extremities of the mine. For two years, the twin objectives of increasing production to full mill capacity and marketing a profitable grade of ore were achieved. Then, in October, 1927, an old enemy—fire—struck, completely destroying the buildings at the Pennsylvania level of the mine. At the termination of the Wells era conflagrations had plagued the company and produced the disastrous financial situation that Livermore had to contend with upon his arrival. In weighing the Smuggler-Union's plight in 1928, Robert Livermore could still be optimistic: ". . . the same spirit of endeavor exists as in former crises, and the best will be made of what at first seemed an overwhelming disaster."[12]

Livermore was whistling in the dark, for as the winter of 1928 approached, whatever optimism he may have possessed faded: "Telluride is getting to be an old story, and to me at least, with a dying industry, and sheep infesting every grazing place from the peaks down, no longer a place of romance."[13] A year previously he had flatly told the directors that unless a hundred-thousand-dollar developmental fund was raised, the Smuggler-Union's days were numbered. In a post-mortem on the Smuggler in 1930, Harold Worcester, a former superintendent, described the crux of the problem: without going into debt for new equipment and construction,

[10] Robert Livermore, "Prolonging the Life of the Smuggler-Union Enterprise," *Engineering and Mining Journal*, CXXXV (March 31, 1928), 1–7.

[11] *Ibid.*, p. 5.

[12] *Ibid.*

[13] Livermore, "Diary," p. 135.

the more productive areas of the mine were impossible to develop. Yet without richer ore, the Smuggler-Union mill could not operate profitably.[14] Here was a Gordian knot that could be severed by only one weapon—more capital. Livermore's search for funds failed. In a reminiscent mood he wrote Worcester complimenting him on his recent article: "No one except those few of us who carried the burden will know of the efforts I made to exhaust every expedient, not only by our operations, but by attempting to enlist other capital, only to have them all fail in the end."

Livermore went on, "The closing paragraphs of your article brought back to my mind one of the saddest parts of the enterprise to me, namely, that I had to ask everyone to stay to the finish, knowing that when the job was done, I had no assurance of providing other work or suitable rewards for their loyalty." As he thought back on those harried months, the only satisfaction that came was from the knowledge that "I had kept the mine going for several years beyond what might have happened, indeed should have happened, had I not taken charge. . . ." After all, Livermore noted, he had gambled "at the risk of reputation, to making the old property pay off its debts, and in doing so had provided a living for our old employees far beyond what they would have had otherwise."[15]

The five years of frustration, moments of hope, and moods of despair terminated on December 6, 1928, when the last miners slowly walked out of the tunnels and down the hill to Telluride. The Smuggler-Union, after half a century of operation, had closed. Livermore might have taken some consolation in the fact that his recommendations for more developmental work and the consolidating of mining claims were accomplished after World War II. The Idarado Company, a subsidiary of the Newmount Mining Corporation, took over the properties and created a flourishing enterprise, in part by following the very suggestions Livermore had made three decades earlier.

Several months before the last car of ore moved down the Smuggler-Union tramway in 1928, Livermore issued a prospectus in conjunction with his partner, W. Spencer Hutchinson, entitled

[14] Harold Worcester, "Closing Days of the Smuggler-Union," *Engineering and Mining Journal*, CXXX (October 23, 1930), 370–374.

[15] Robert Livermore to Harold S. Worcester, November 4, 1930. Worcester Collection, Western History Research Center, University of Wyoming.

"Proposal for the Incorporation of a Mining Investment and Exploration Company." The plan was simply for an investment company to acquire stock in old reliable firms and investigate promising new mines. The firm of Hutchinson and Livermore would serve as consulting engineers. From this beginning developed the North American Mines, Incorporated, in which Livermore was associated with old family friends, Quincy Shaw and Palmer Gavit.

Though it was an arduous task to attract capital "simply on the promise that we could use it well," Livermore had amazing success. By January, 1929, North American Mines had a capitalization of six hundred-thousand dollars—all paid up."[16] At the end of 1929, he laconically remarked in his diary, "Much philosophy is needed these days."[17] The next two or three years were spent in investigation, appraising, and discarding prospective mineral acquisitions, with the heaviest attention being given to properties in Baja California, a tungsten mine south of Tucson, and Nevada copper.

Baja California fascinated Livermore with its scenery, picturesque towns, and flocks of children. From February 22 to March 11, 1930, he wandered in company with friends over countryside, examining one prospective property after another. Though he was dubious about investing in most of what he saw, Livermore did think the country might have possibilities: "One impression I get of this country is its virgin quality compared to the 'States.' Any quantity of promising areas and prospects hardly scratched. It must have been like this in Arizona 40 years ago or more."[18]

After his return in the spring of 1930, Livermore organized another trip to Baja California that summer, chiefly because several of his Boston friends and investors wanted an opportunity to look over the properties. This time they definitely decided to invest. The following year was spent in attempting to obtain titles, negotiate in an explosive political climate, and making short trips to Arizona and California. The last ray of hope vanished in Baja California by mid-1931. On his last trip, North American mines employed the well-known copper geologist Ira B. Joralemon, who turned down the mine on which they had begun exploratory work. Livermore lamented, "All the news from the mine was bad and this made the trip not

16 Livermore, "Diary," p. 136.
17 Ibid., p. 141.
18 Ibid., p. 148.

a joyful one for me as I could see all my hopes of fortune vanishing."[19] Nor was Livermore any happier with the intelligence that the tungsten enterprise in Arizona was also failing.

The next two years were occupied by investigating mines in New Mexico, opening an old gold district near Fredericksburg, Virginia, and taking over the Tincup mine near Kingman, Arizona. In 1934, Livermore took a trip to the Yukon to examine another promising prospect. Writing from the *S.S. Casca* on the Yukon River on August 9, he commented: "Skagway is a dreary single street reminisicent of the western mining towns now nearly deserted. A narrow guage rail took us through White Pass, the old trail of '98 much like one of the Colorado passes with high granite mountains and alpine shrubs."[20] Another trip, another disappointment; even before his trip down the Yukon was finished, he noted, "My present impressions are what a hell of a long way we have come for a mine."[21]

Where there is another mine, there is another hope. Livermore took over the Gold Belt mine in British Columbia in early 1935. It, too, proved an illusory anticipation. The following decade, he spent consulting with North American Mines and with Calumet and Hecla, being vice-president in both firms.

In 1947, he retired to his home, "Boxfields," in Boxford, Massachusetts; death came at the age of eighty-three, September 26, 1959.[22] His wife Gwendolen died on November 27, 1965. Today her son, Robert Livermore, Jr., and a daughter, Mrs. Gwendolen Woodward, live in Massachusetts. Another daughter, Mrs. Cecily Beal, resides in Cumberland Foreside, Maine. Twenty-four years before his death, Robert Livermore had summed up both the fascination of his career and the philosophy of his life, in a talk before his friends in the Boston section of the American Institute of Mining Engineers:

> One of the things which makes life interesting is contrast. My partner and I loved contrasts; our greatest delight was after months of hard work at high altitudes, to start horseback, clad in digging clothes, descend to town,—bath and a shave,—then hie us Eastward, achieving more luxury with every step. At

---

[19] *Ibid.*, p. 162.
[20] *Ibid.*, p. 172.
[21] *Ibid.*
[22] Boston Sunday *Herald*, September 27, 1959.

Chicago, we doffed Stetsons and purchased, "2 hard hats." In New York, nothing but a "topper" and all that goes with it would go. . . .

Years later, I remember another contrast. On a visit to New York, as usual seeking solace from underground grime, clad in Broadway togs, we received word to examine a new strike 100 miles south of Hudson Bay;—three days later, in furs, in a dog team, under Northern lights wondering whether we would get somewhere before we froze to death. Most of us know these contrasts of a mining life, which is one of the things that make it interesting.

I always remember a book I read once beginning: "I have had a glorious life." He was a Mining Engineer who wrote it, and I said: "Check, so have I! and I expect to continue having one." [23]

[23] "Twenty Seven Years of Work and Fun in Mining," April 1, 1935, MS, Livermore Collection, Western History Research Center, University of Wyoming.